The Intellectual Appropriation of Technology

The Intellectual Appropriation of Technology
Discourses on Modernity, 1900–1939

edited by Mikael Hård and Andrew Jamison

The MIT Press
Cambridge, Massachusetts
London, England

© 1998 Massachusetts Institute of Technology

All rights reserved. No part of this book may be reproduced in any form by
any electronic or mechanical means (including photocopying,
recording, or information storage and retrieval) without permission in
writing from the publisher.

Set in New Baskerville by Crane Composition, Inc.
Printed and bound in the United States of America.

Library of Congress Cataloging-in-Publication Data

The intellectual appropriation of technology: discourses on modernity,
1900–1939 / edited by Mikael Hård and Andrew Jamison.
    p. cm.
    Includes bibliographical references and index.
    ISBN 0-262-08268-3 (hardcover: alk. paper).—ISBN 0-262-58166-3 (pbk.:
alk. paper)
    1. Technology—Social aspects—History—20th century. I. Hård,
Mikael. II. Jamison, Andrew.
T14.5.I577 1998
303.48′3—dc21   98-18003
                                                                      CIP

Contents

Acknowledgments

This book has been written within a research project on Technology and Ideology that was supported by the Humanistisk-samhällsvetenskapliga forskningsrådet (HSFR) from 1989 to 1994. The project was based at the Department of Theory of Science at Gothenburg University. We would like to thank both the research council and the department for their support and, of course, our contributors for their help in bringing the work to fruition.

More specifically, we would like to express our appreciation to everyone who commented on the various chapters at seminars and conferences through the many years of research and writing. Particularly important were the first round of discussions, in Gothenburg in 1990, when the general contours of the book were shaped, and the final round, in Lund in 1994, under the auspices of the Research Policy Institute at the University of Lund, when the manuscript had begun to take coherent form. We would especially like to thank Ron Aminzade, Håkon With Andersen, Ron Eyerman, and Orvar Löfgren for their participation in the Lund workshop, and Martin Kylhammar for his continuous interaction with the project.

A somewhat different version of Dick van Lente's chapter has appeared in the *International Review of Social History,* and we appreciate being able to publish the present version.

The Intellectual Appropriation of Technology

1

Conceptual Framework: Technology Debates as Appropriation Processes
Mikael Hård and Andrew Jamison

Modernity and Its Discontents

The twentieth century has been marked by a schizophrenic relation to technological development. The technological optimism that served as the main story line through most of the nineteenth century in both Europe and North America was transformed, in the decades surrounding World War I, into a widely felt spirit of disillusionment. In the words of Lewis Mumford (1982: 377), "The 1914 war had left my generation, even those who had taken no active part in it, in a state of shock: the unthinkable had happened, and from now on nothing would be quite unthinkable." The benefits and progressive features of modern, complex technology have since then been countered at seemingly every turn by questions, problems, and critical reactions. Should human societies seek to make technical systems and their products "appropriate" by assimilating them into existing values and organizations (Schumacher 1973), or should the values and structures of society be adapted to the inherent "logic" of *la Technique* (Ellul 1964)? Can societies find ways to steer technological development into acceptable directions (Rip et al. 1995), or does technology inevitably determine social changes (Smith and Marx 1994)?

In general terms, these questions have become ever more difficult to answer. Out of disillusion has grown, in the course of the twentieth century, a highly variegated set of responses to the challenges of modern technology. To deal with its discontents, humanity has put technology in its place, no longer imagining the machine as an intrinsic source of progress and improvement but endowing it instead with contradictions, tensions, and dichotomies. The machine has filled the twentieth-century mind with *both* fascination and foreboding, and which feeling dominates has come to depend on the particular technology under reflection and on the particular time and place of assessment. In other words, how a technological project is debated and judged, and how it is ultimately understood, has come to be specified and particularized—in short, situated in context (Cutcliffe and Post 1989).

This book explores the ways in which that contextual understanding emerged in the first four decades of the twentieth century, primarily among intellectuals but also among other users of technology. What Friedrich Dessauer (1958) later called the "struggle about technology" was first articulated explicitly, we contend, in this period. Facing a world of tanks and assembly lines, and using telephones and automobiles, "modern" men and women were forced to differentiate the uses of technology from the abuses. The proliferation of the machine into ever more areas of social and economic life led to a need to interpret its meanings in a much more comprehensive way than in the past. The symbolic significance of technology became subject to contestation when, through new applications of science and engineering, machinery began to fundamentally alter the reality of war and "class struggle." As its material manifestations were constructing what Kern (1983) has termed a new culture of time and space, technology became important to historians and philosophers as well as to the newfangled social scientists. In this period, technology not only became an integral part of everyday life but was also given a place in the intellectual universe. It was in this period that Lewis Mumford wrote *Technics and Civilization* (1934), a work that perhaps more than any other has shaped the historical conceptualization of technological change. It was then that Dessauer, Peter Engelmeier, José Ortega y Gasset, and others began to articulate a modern philosophy of technology (Mitcham 1994). And it was in these years that Max Weber, William Ogburn, Thorstein Veblen, and Joseph Schumpeter raised topics that social scientists have struggled with ever since (Jamison 1989).

In the first four decades of the twentieth century, writers and publicists began to discuss what might be termed the civilizational aspects of machinery. Modern technology could no longer be rejected by escaping to a Thoreauvian Walden Pond, but its social and cultural consequences could certainly be subjects of debate and reform. These early-twentieth-century commentators were no longer confronting technology as a historically new phenomenon; they were, rather, dealing with a series of systemic transformations in which science-based technologies played an important, even a central role (Hughes 1989). Their solutions to the social and cultural challenges raised by the machine system are still formative for our contemporary responses; that is, we continue to deal with technology, to a significant extent, by making use of conceptual frameworks, policy structures, and social and institutional contexts that were established between 1900 and 1940.

By returning to these early technology debates, we aim to contribute to an understanding of our own era's ambivalent relation to technology. Analyses of technology debates, or of ideological positions on technical matters, have often been one-dimensional. "Positive utopians" have been placed in opposition to "negative critics" (Huber 1989); the scientific-technological culture has been juxtaposed with the humanist culture (Snow 1959); progressives have been treated as the natural opponents of conservatives. When dealing with specific technological problems, such as the future use of nuclear power or the ethical acceptability of the Human Genome Project, we might indeed be able to analyze debates in this way, in terms of pros and cons; however, when discussing broader interrelations between technology and society, we require a more differentiated analysis. It is seldom appropriate to say that an individual or a social group is opposed to all kinds of modern technology and that others are unreflectively positive. Greenpeace activists struggle against many aspects of modern industry, but they nevertheless employ up-to-date communications and information technology in their campaigns (Eyerman and Jamison 1989).

Our way of coming to grips with technology debates is to distinguish the nuances of various intellectual positions in a less polarized and more comprehensive way than is common among historians and social scientists and to place debates about technological development within their political contexts and intellectual traditions. Instead of focusing simply on uncovering dichotomies between proponents and critics, we will show how these dichotomies worked themselves out differently in different national and organizational settings, both in terms of the cultural resources and traditions that were mobilized and in regard to the international influences that were significant. In this manner, we want to contextualize the discourses about technology, but we also want to relate what was said then to the problems that confront us now. For example, "Fordism" and "Taylorism" took on quite distinct connotations in different European countries, depending on their industrial history and their particular ties to the United States (Nolan 1994).

Our concern with history is informed by contemporary problems and an ambition to uncover some of the roots of present ways of dealing with and discussing technology; but we are not attempting to read into, or impose upon, history events that never took place or positions that never were taken. Our present situations and discourses—academic and otherwise—lead us to ask of the historical record certain questions that until quite recently have rarely been asked. Technological development in our day has become so problematic, and so filled with controversy, that it has become a matter of some urgency to look at how our conceptions of technology have been constituted.

Although contemporary problems have helped us to select our topic of investigation, we see our task as primarily historical. We want to amend the standard historical accounts by showing that responses to technology were far more central to the making of modernity than is usually recognized. By analyzing the ways in which intellectuals appropriated technology into "discourses of modernity," we want to indicate that the ideas about technology that were articulated during the early part of the twentieth century have been influential for and formative of the modernization process itself. Our goal, therefore, is to provide historical material that can be of interest to sociologists, political scientists, and philosophers as well as to historians.

Most historical accounts tend to neglect intellectual debates about technological development, and for this reason the role of these debates in the making of modern society has not received the attention it deserves. The efforts made by intellectuals to come to grips with the challenges raised by modern technology tend to fall between historiographic stools. In the history of technology, there has been a strong tendency to delimit attention to particular debates about particular technologies; it is as if the historians of technical change had not been able to deal with the general, more "theoretical" issues. The historiography of technology has been characterized by case studies and specific instances; longer-term movements or processes have been left to the popularizers, who are often looked down upon by practicing historians. The tendency has been to subdivide the social appropriation of technology into separate historical specialties within the historical profession: economic history, where the productive aspects have been analyzed (Landes 1969); literary history, where the responses of creative writers have been elucidated (Marx 1964); social history, where institutional and popular processes of assimilation have been investigated (Nye 1991, 1994); and intellectual history, where philosophers and other public intellectuals have been taken up (Herf 1984). For their part, most philosophers of technology have tended to search for universal notions of instrumental rationality or technological "lifeworlds" rather than examining specific historical contexts (Mitcham 1994). Others use history primarily as a garden from which to pick ideas that support their own theories (Feenberg 1991; Rothenberg 1993).

Among cultural historians, and in cultural studies more generally, intellectual responses to technology have, until recently, rarely intruded into the rarefied air of artistic, literary, and philosophical expression. Latter-day humanists have had little time for and apparently less interest in technology, and even though the "question of modern technology" loomed large in the early part of the century—before C. P. Snow's notorious two cultures were placed firmly into the modern intellectual consciousness (and educational curricula)—systematic scrutiny of their predecessors' debates has not interested many intellectual historians. The tendency has been to look at the intellectual responses to modernity in an overly specific, delimited, and individual fashion, usually by writing biographies or descriptive surveys (Tichi 1987; Blake 1990), or to leap into grand theoretical reflection about the modern and, increasingly, postmodern "condition" (Lyotard 1984; Jameson 1991). We have thus a voluminous case literature of individual biographies and circumscribed historical periods, and a growing theoretical literature of discourses and thought structures and "archaeologies of mind" (Foucault 1966), but the comparative and synthetic works that are needed to fill the gap between detail and discourse are few and far between.

The Historical Setting

As already mentioned, the debates about technology did not begin in the twentieth century, but they differed in important ways from those that had come before. There were two distinct periods of debate in the nineteenth century, which can be seen as responses to the first and second "long waves" of industrialization (Freeman 1984). In the early nineteenth century, the new mechanical instruments of production encountered direct opposition from the machine-breaking followers of Ned Ludd and Captain Swing (Pearson 1979). Workers strategically destroyed the machinery that they saw as responsible for their "redundancy." The mythical Ned Ludd called on his armies to confront technology head on, and the British army was brought out in force to impose the new mechanical order. Mechanization was rejected by poets too: Blake's "dark satanic mills" linked mechanization to the devil, while Keats and Byron escaped from the mechanical world to carve out an aesthetic world, to counter the machine with beauty. Many were the romantics, in art, music, and everyday life as well as in literature, who turned their backs on industrial society to gain inspiration from the wilderness or from the ideals of earlier, pre-industrial epochs (Roszak 1972). Most significant, perhaps, were the literary experiment of Mary Shelley, who imagined the industrial paradigm in the form of the monster constructed by her mad Doctor Frankenstein, and the experiment with life that the young Henry David Thoreau conducted in the 1840s on the shores of Walden Pond. Thoreau built his own house, grew his own food, and used the opportunity to reflect on the underlying meanings of the emerging industrial order (Marx 1988).

The romantic revolt against the machine—the personal, direct opposition that was manifested in various forms in the first half of the nineteenth century—was largely overtaken, or at least made problematic, by the course that technological development was to take. Eventually, humanists and cultural critics articulated a more sophisticated response. Thomas Carlyle's and Friedrich Nietzsche's opposition to machinery and industrialization helped form the philosophical and critical traditions in several European countries (Williams 1958). At the same time, proponents of technology sharpened their arguments and gave machinery a central place in the emergent social sciences, as well as in the procedures of industrial management and organization. The "machinery question" shook British society and heavily influenced the development of political economy. It has been suggested that economics as an academic discipline and the labor movement as a political force were both formed largely as social responses to mechanization (Berg 1980). But in responding to mechanization, they also served to change the nature of the critique. From being a force of the devil and of mankind's darkest emotions, technology became a fundamental economic factor, or productive force, for "scientific" socialists like Marx and Engels, and a potential source of new forms of collective action for others, including William Morris in Britain and Edward Bellamy in the United States. In the latter half of the nineteenth century the rejection of technology was supplemented by more constructive kinds of appropriation, from the British Arts and Crafts movement, with its industrial mobilization of traditional images and techniques, to the literary genre of science fiction, with its visionary prognostications reflecting on the human implications of scientific and technological progress. Technological pessimism, on the other hand, developed, especially in the crazed mind of Nietzsche, into an all-encompassing philosophical conception. In the notion of the superman, and in the diagnosis of the underlying contradictory emotions that technology-based progress represented, Nietzsche disclosed, and ultimately suffered from, the schizophrenia of technological society. In the twentieth century, his view would gain new and increasingly tragic force (Ezrahi et al. 1995).

The pessimists were in a minority position through most of the nineteenth century, although their influence differed from country to country. Later commentators have, for instance, argued that there was a distinctly more "anti-industrial" spirit in Britain than in North America and Germany (Wiener 1981). The main currents of social, political, and cultural thought, however, incorporated technology within the established social and cultural frameworks, while the engineers sought to make technology itself into a "science" and to adapt scientific rituals and titles to their more practical concerns. The positivist philosophy that had emerged from the Ecole Polytechnique in Paris, one of the world's first technological universities, accompanied the spread of the machine across Europe and North America. Science and technology represented progress; they were the central defining characteristic of the age. And many intellectuals paid homage to the new forces of production by placing them at the center of their theories and analyses. They sought to infuse a technological fascination and interest into economics, philosophy, literature, and history. From August Comte to August Strindberg, the enthusiasts of technological development urged on their fellow intellectuals a greater awareness of technology and of its importance for the shaping of the modern, Western industrial order. In the process, they initiated new discourses about technology—in particular, those of Karl Marx (in social and economic theory), Herbert Spencer (in sociology), Jules Verne (in literature), John Ruskin (in art history), and Henry Adams and Frederick Turner (in history).

This volume addresses these nineteenth-century debates only in passing, insofar as they form the background and the roots of the discursive frameworks in which twentieth-century intellectuals came to debate the meanings of modern technology. Our focus is on the discussions about technology during the period that Peter Wagner calls the "first crisis of modernity," when the subject of debate was the project, rather than the products, of technological change (Wagner 1994). Not only did economic liberalism come under attack; so did the ideas of democracy and science. The growing power of the working class opened the way for far-reaching collective initiatives and ideas, and political instability opened up the possibility for radical authoritarian solutions. This was not only the period of what Rudolf Hilferding called "organized capitalism," with the state taking on a range of new functions and activities; more generally, it was the period of organized modernity. While cartels and socialization took on importance in the economic arena, collective ideals such as Rudolf Kjellén's "people's home" and Adolf Hitler's Third Reich gained strength in the political arena.

In this context, the "machinery question" was replaced by the "struggle of technology" (Dessauer 1958). The debaters of the early twentieth century were concerned not only with the consequences of mechanized production but also with the mechanization of life as a whole. Most famous in this regard is Max Weber's analysis of the Occidental rationalization process that brings with it a loss of freedom and meaning in the name of reason. The role of technology in this process is, metaphorically, to provide the bars of the "iron cage," and, more concretely, to simplify control and subordination of what the German philosophers Edmund Husserl and Jürgen Habermas have called the lifeworld. As Weber asked rhetorically in his famous lecture "Science as a Vocation" (1919):

Does [intellectualist rationalization] mean that we . . . have a greater knowledge of the conditions of life under which we exist than has an American Indian or a Hottentot? Hardly. Unless he is a physicist, one who rides on the streetcar has no idea how the car happened to get into motion. And he does not need to know. . . . The savage knows incomparably more about his tools. . . . The increasing intellectualization and rationalization do not, therefore, indicate an increased and general knowledge of the conditions under which one lives.

The reactions to the growing importance of technology were split. On the one hand, the "revolt against positivism" (Hughes 1958) encouraged the development of a philosophical critique of technological civilization, particularly in Germany but also in Italy and Spain. José Ortega y Gassett, a Spanish philosopher who often addressed the question of technology in his writings, pointed to the ultimate meaninglessness of twentieth-century life: "Just because of its promise of unlimited possibilities technology is an empty form like the most formalistic logic and is unable to determine the content of life. That is why our time, being the most intensely technical, is also the emptiest in human history." (Ortega y Gassett 1958: 151) Ortega was not opposed to technology (indeed, he claimed that opposition had become an impossibility in a world that had grown dependent on technical achievements); however, he questioned, as did many philosophers, the peculiar amorality of the machine and the artifactual society it had brought into being.

On the other hand, there were intellectuals who devoted their energies to devising new modes of accommodation to a mechanical civilization. In all of the countries discussed in this book, there were discussions of technology policy, regulation, control, and assessment—some 50 years before the establishment of the Office of Technology Assessment in the United States. Early in the twentieth century, it was dawning on growing numbers of intellectuals that new social mechanisms and institutional frameworks were needed to bring order and control to a technology-based society. Whether technocratic, pragmatist, or based on social engineering, schemes for managing the technological civilization took on an increasing importance.

This book thus focuses on the period of transition between two phases of industrialization: the first 40 years of the twentieth century. World War I and the economic depression of the early 1930s constitute two discernible turning points in all the countries under consideration here. But there are, as we will try to show, also points of reorientation that are country-specific and without global ramifications.

World War I is usually seen as the Great Divide in the intellectual history of modernity. Before the war, it is often said, most people were optimistic and believed in the future. They regarded science and modern technology as the prime movers of prosperity and were primarily positive about what industry had wrought. With the war all this disappeared. Optimism died with the soldiers in the trenches, and the belief in the blessings of science and technology was killed by mustard gas and tanks. We do not want to claim that the war had no impact on the intellectual scene; certainly it did. We do, however, suggest that its consequences were not so simple and universal as the standard accounts would have it. The cruelties exposed by the war did challenge both the intellectuals and the general public, but their reactions were mixed. While some writers were truly appalled by the reports from the front, others saw the war as a healthy purge that would be followed by a new beginning. Whereas some authors found the war to have demonstrated the bankruptcy of Western civilization, others thought it would lead to the appropriation of modern technology on a national basis (Jünger 1929).

The reactions to the war also differed among nations. Among the countries under discussion here, they were most dramatic in Germany. Not only did Germany lose the war; in addition, it was forced to pay an enormous war indemnity, and its political system was seriously affected. The Kaiser had to resign, and the bourgeois state was threatened by socialist movements. In the United States, the war led to a sense of disillusionment among many intellectuals and to a rejection of the materialism that was seen as characteristic of the modern "wasteland," which came to be depicted as a deterioration of moral values by Joseph Wood Krutch, T. S. Eliot, and other literary intellectuals; however, not until the stock market crashed in 1929 did the disillusionment become widespread. The breakup of the union between Norway and Sweden in 1905, long before World War I, signaled the need for a similar reorientation in the latter nation. It became clear to the Swedes that their country was no longer as geographically large and impressive as it had been, and in their search for a new foundation they turned to modern industry. Technological development would transform Sweden into a great power in much the same way that military technology had provided the basis for Sweden's great-power ambitions in the seventeenth century. In other words, it was in particular years—1905 in Sweden, 1918 in Germany, and 1929 in the United States—that the challenges of modern technology led to a radical symbolic shift and a new beginning. The shift did not only take place on the political, military, and economic levels; it also did so on the discursive level. In Sweden the new path was couched in terms like "a new period as a great power"; in Germany there was a discussion about how to organize "the new state" and "the new economy," and in the United States there emerged Roosevelt's "New Deal."

The new beginnings were not altogether new, however. Our thesis is that each country met its challenge by reverting in part to culturally specific ideas. The term "a new period as a great power" implied that Sweden should become as great as it had been in the seventeenth century. Older progressivist ideals were reinvented under a more palatable disguise in the United States. And in the German setting recourse was made both to traditional ideas of a unified, strong, and paternalist state and to the classical ideals of *Bildung* and *Kultur*. By returning to longstanding national traditions and conventions, each country's intellectuals tried to come to grips with a challenging situation in a nationally specific way.

The Comparative Perspective

A survey of some of the world's leading historians of technology has shown that they regard national comparisons to be one of the most pressing research tasks in their discipline (Hård 1989). The seeds that Thorstein Veblen (1915) planted in his comparison between the German and British roads to industrialism have only recently begun to bear fruit. Veblen's main concern—formulated in the shadow of the war—was to show how different "modes of habituation" drove "the instinct of workmanship" in different directions. Comparative studies are nowadays an important method for those who do not believe in the autonomous character of technology and/or the objective nature of scientific knowledge. They effectively remind us that the direction and the content of technological and scientific developments are not necessarily globally uniform (Hård 1994a), and they teach us that we must treat technology and science as culturally dependent variables.

One may, of course, question whether the nation is always the most suitable level of comparison (Radkau 1989). Some historians have chosen to discuss regional or municipal patterns (Hughes 1983), and others have focused their research on company particularities (Chandler 1990; Kenngott 1990). In several instances it does, however, seem perfectly justifiable to discuss science and technology along national lines. Since the nation-state is still the central unit for the formulation and implementation of research and development policies, it makes sense to analyze policy questions along national lines (Jamison 1991; Nelson 1993). Similarly, geographical and institutional factors certainly affect which technologies are developed and chosen in various countries. For example, Colleen Dunlavy (1994) has fruitfully analyzed "national styles of railroad technology" in the United States and Prussia in terms of different configurations of interests and organizations. Hans-Liudger Dienel (1995) has, similarly, compared the different ways in which refrigeration technology was developed and implemented in the American and German settings, owing to different engineering traditions and production systems. The economic historians H. J. Habbakuk (1967) and Sidney Pollard (1981) have attempted to show that national economic patterns influence the content and structure of technology.

The present book does not directly deal with the formulation of research and development policies or with the construction of national systems of innovation. It focuses instead on the discursive, conceptual, and ideological frameworks within which such processes take place. The design and the use of even the simplest artifact require that it occupy not only a physical but also a mental space (Dierkes et al. 1996). For instance, W. Bernard Carlson has claimed in an article on the early history of the American film industry that the different ways in which Thomas Edison and his competitors constructed their movie-making systems depended on their having different "frames of meaning." These frames include inventors' and entrepreneurs' "assumptions about who will use a technology and the meanings users might assign to it" and "thus directly link the inventor's unique artifact with larger social or cultural values" (Carlson 1992: 177).

Our book focuses on these "larger" values, assumptions, and norms. Although there is now a voluminous literature on the modern project, the material components of modernity are often bracketed out of the analysis (Bauman 1991; Giddens 1990; Habermas 1987). Unlike many social theorists and other writers on modernity, we want to focus attention on the meanings attributed to modern technology in different countries. The book investigates some of the ways in which visions of technology are shaped by national intellectual traditions. Considering the close connections between assumptions and language, we also examine how distinctive conceptual frameworks develop in different linguistic contexts. To some extent, our effort can be regarded as an attempt to unravel Oswald Spengler's (1931: 83) suggestion that modern industry was valued most highly in "the practically oriented America" and much less so in "Germany, 'the country of writers and thinkers.'"

"Contextualism" has become a catchword in the interdisciplinary field of science and technology studies in the last decade or two. John Staudenmaier (1985: 139) has observed that historians of technology increasingly assume "that every technology is local-specific"—that "at every stage of its development, from invention to obsolescence, a technology can be explained historically only in terms of relationships with its particular [cultural] ambiance." Similarly, Merritt Roe Smith and Steven Reber (1989) have pointed out that pure internalism is becoming an ever more peripheral position. The same trend is also visible in the sociology of technology, where since 1980 various proponents of social shaping and construction have explored the contingent and local character of technological development (MacKenzie and Wajcman 1985; Bijker 1995b).

Intellectual Traditions and Discursive Frameworks

Most of the characters in this book are intellectuals, people who make their living by using the pen or the lectern. Some of them derive their authority from their office. For instance, the German "mandarins" discussed in chapter 3 had what Pierre Bourdieu calls "symbolic capital" because they held positions as governmentally employed professors. Other intellectuals lived more precariously. For example, many of the

American writers encountered in chapter 4 were dependent on their books' and articles' finding an audience in a commercial market. We follow the sociologist Ron Eyerman in suggesting that the social position of intellectuals is not the same in all societies and at all times: "Intellectual is . . . understood as a situated social practice, not a fixed quality, and intellectuals by the specific social relations which constitute that practice" (Eyerman 1994: 6).

Unlike Karl Mannheim (1936), we do not treat intellectuals as a "classless stratum" or as equivalent to a "socially unattached intelligentsia." Intellectuals may be found in all social groups, and an intellectual's views are often affected by the social position occupied by his or her group. Following the historian of ideas Sven-Eric Liedman (1986), we could say that an intellectual's ideology is dependent on, although not determined by, his *utsiktspunkt*—the point from which he views society. The same could be said about generations. Like all people, intellectuals are formed not only by their social positions but also by the times in which they live. In particular, their points of view seem to be influenced by experience gained in youth. Detlev Peukert (1987: 25–31) has developed the thesis that German intellectuals and politicians born between 1840 and 1900 can be divided into four generations, each of which had a specific experience that affected their views and practices. For example, involvement in World War I contributed to a strong feeling of revenge in Adolf Hitler, born in 1889.

Intellectuals contribute to the formulation of ideas and to the manifestation of consciousness in a society. They are important in framing the discourses in which various social groups express themselves. Eyerman emphasizes that in doing so they usually bring a historical perspective into the pressing daily problems of their society. Intellectuals make use of history in two ways. They provide society with a history, without which its members would have substantial difficulty orienting themselves, but they also perpetuate and often take part in the "invention of tradition" (Hobsbawm and Ranger 1985). Or, as Eyerman (1994: 16) puts it, "tradition refers to the cultural resources upon which new generations of intellectuals draw to reinvent their role in new contexts." In the ensuing chapters we will encounter many instances in which intellectuals met new challenges by making recourse to assumptions and values that their contemporaries regarded as "traditional" and "cultural." In the Swedish case, for instance, the traditional, centuries-old dream of Sweden as a great power provided intellectuals of various political colors with powerful arguments in a new and serious situation. As David Gross (1992) has pointed out, traditions can play progressive as well as conservative roles. Traditions are important not only in the shaping of the intellectual role but also in the formation of larger intellectual discourses.

In this book we introduce the concept of a discursive framework to indicate the broader terms of reference within which intellectual positions are formed. Intellectuals juxtapose tradition and context by applying cultural resources that have been shaped by deep-seated historical processes to contemporary situations and issues. Their thoughts and expressions are structured by their discursive frameworks but are by no means determined by them. This concept builds on the work of Quentin Skinner, who has portrayed political ideology formation as a creative "manipulation of conventions" (Tully 1988). In this Wittgensteinian world of language games, an idea is never completely the product of an individual mind; rather, it is articulated within a linguistic framework that imposes restrictions on thoughts, actions, and utterances. Intellectual innovators, such as Skinner's Machiavelli, redefine the restrictions in response to political problems of the day. For us, discursive framework is a concept that allows the common conventions of groups of intellectuals to be analyzed in both a contextual and a historical perspective. The discursive framework provides a vocabulary and a "way of seeing" that is shared by a larger collectivity (be it national, regional, or generational).

Analyses in terms of discursive framework make national comparisons particularly interesting as a test bed. It may be assumed that a common language, partly shared cultural values, a national education system, and common debate forums contribute to the nationalizing of concepts, of expressions, and of (by implication) thoughts. This book attempts to show how reactions to modern technology were strongly shaped by nationally distinctive discursive frameworks. In Germany that framework was based on traditional concepts of *Bildung* and *Kultur*: many commentators, both technical and literary, came to see their task as redefining the classical intellectual ideals of cultivation and evaluating them critically in relation to modern technology. In the post–World War I context, they saw the challenge of the times as a need to reaffirm and redefine the German national identity. In the United States, a waning "republicanism" that had endowed technical and industrial development, through much of the nineteenth century, with a set of egalitarian and artisan meanings was called upon by some intellectuals to provide a "usable past" for the present, while others sought to create from similar materials an ideology of Americanism that would show the rest of the world how to deal with modern technology. In Sweden, technology was welcomed in most quarters as a path to modernization and, through linkages to historical symbols of greatness and power, was given the role of restoring a sense of national pride. When facing what they found to be a globally uniform technology, intellectuals in various countries thus mobilized national cultural resources in response. However, the intention of most intellectuals was not to block or reject technology but rather to find means by which they could accept it. The book suggests that intellectuals tried to ameliorate the incorporation of modern technology by finding a place for it in one or another discursive framework. It is this process that we call intellectual appropriation. It implies that men and women of the early twentieth century tried to come to grips with modern technology by taking it out of a frightening and strange world of mechanized relations and placing it in a more familiar world of tradition and linguistic convention. As in Martha Banta's discussion about how the term "efficiency" was appropriated by writers of different ideological dispositions, we will show how intellectuals shaped the meaning of modern technology in terms that fitted their respective cultural heritages (Banta 1993). The goal usually was either to assimilate technology into the existing culture or to adjust culture to the intrinsic demands posed by technology.

Appropriation processes have largely been discussed by sociologists (Mackay and Gillespie 1992), although the recent work of Banta and Nolan indicates that there is a growing interest among historians as well. It has been shown how users try to domesticate and incorporate technologies by making them fit into existing lifeworlds (Lie and Sørensen 1996). Although a cognitive dimension may be found in some of these studies, their focus is on habitual action. This book attempts to broaden that focus and to make visible the various discursive means by which intellectuals frame society's understanding of novelty, and how they thereby enable society to accept it and accommodate to it in an active manner.

2

Theoretical Perspectives: Culture as a Resource for Technological Change

Aant Elzinga

Discourses and Political Urgency

As a new century approaches, several issues that have been with us for at least 100 years stand out and cry for political action as never before. Prominent among these is the question of managing technological change and use without jeopardizing quality of life. In our day this question has, in part, become couched in a discourse on "sustainable development." Human-made changes in the chemical composition of the atmosphere and the stratosphere have now reached a point where, thanks to science, it is becoming clear to increasing numbers of people that unfettered economic growth is not viable, if it ever was. *Our Common Future* (1987), the report of the World Commission on Environment and Development, is usually referred to as the benchmark by which further development must take into account effects on the multifaceted ecological system of the planet, locally and globally.

In a recent policy article published in *Science*, Kenneth Arrow and colleagues observe: "The environmental resource base upon which all economic activity ultimately depends includes ecological systems that produce a wide variety of services. This resource base is finite. Furthermore, imprudent use of the environmental resource base may irreversibly reduce the capacity for generating material production in future. All this implies that there are limits to the carrying capacity of the planet." (Arrow et al. 1995, p. 520) Carrying capacity, we are told, is not fixed, static, or simple; it is contingent on technologies, preferences, the structure of production and consumption, and the changing state of interactions between the physical and the biotic environment.

Seen in historical retrospect, this is only the latest, albeit a very acute, formulation of a continuing concern over the modernist mode of development. As Peter Wagner suggests in chapter 9, such concerns signify a rejection of the classical liberal doctrine that emerged in the nineteenth century, according to which economic growth and technological development are seen as simply broadening the scope of individual human action and control over nature. By the beginning of the twentieth century, the first insights had accumulated to suggest that technological development and its effects must be understood in a more complex way. Socialist movements called for total overhaul of reigning economic structures, and among liberals and cultural conservatives there was a sense that unfettered growth undermined the very freedom it was supposed to facilitate, at least at the level of individual creativity and culture. Here, industrial and technological growth was seen as a source of standardization, routinization, and regimentation of working life, narrowing of cultural diversity, and erosion of traditional values. With the advent of oligopolistic and managerial capitalism, then, the idea grew that economic and technological change might have to be regulated—if not voluntarily, then through the state. Alongside the perspective of classical liberalism, what Wagner calls the ideal of "organized modernity" emerged as a solution to the crisis and demise of the classical view.

According to the new liberal view, further technological development was no longer a matter only of the broadening of the scope of individual action potential. Increasingly, it was a question of shaping future physical and social spaces. This prompted debate over the relationship between responsibility for the collective good, say at a national level, and local individual freedoms of action. Technological change could mean both new opportunities and new threats, even in such infrastructural areas as transportation and communication. Here we must confront large technological systems that combine to tie societies together, internally and across national boundaries (Bell 1993). Taken by themselves they seem to exemplify modes of means-ends rationality that become increasingly focused; but this is at the cost of losing sight of alternatives that might have been, and with it blindness to their underlying rationale, which from other points of view could be seen as irrational.

Enlightened economic and political elites who found themselves at the helm of these transformative processes spoke of regulation in the name of what they held to be the collective good. As Mikael Hård points out in chapter 3, if these processes are examined more closely, it seems clear that they benefited these elites and the classes from which they came more than they benefited the larger and less privileged sectors of the population. Hence, it is not strange to see opposition to technological change framed in various guises. At one end of the spectrum of positions one can see the notion that the system of governance should be changed; at the other end one can see the view that further technological change should be stopped, since it seems to cause more human suffering, grief, anxiety, and rootlessness than it alleviates. In particular one could point to the erosion of traditional ways of living and of values, to cultural disorientation, to existential anxiety, and to social anomie.

The essays in this book explore various responses to this problem of technological change as it was perceived in the early decades of the twentieth century. We focus particularly on the years following the shock of World War I. A variety of positions have been reconstructed, among them the idea that it is possible to regulate economic and technological change under capitalism provided that there are institutional arrangements that can offer foresight and some modicum of technological assessment. Rudiments of this idea were present in the writings of the German industrialist Walther Rathenau, who, as Hård argues in chapter 3, insisted that the state should intervene as a neutral conciliatory agency to bridge gaps and dissolve conflict between organized labor and capital. In the United States, as Andrew Jamison shows in chapter 4, similar ideas emerged from different cultural settings, leading to proposals with different ideological and political accents. The pragmatic school of John Dewey, for example, argued that new institutions were needed to infuse science and technological development with a new morality that was better equipped to deal with the consequences of modern technological developments. The New Deal politics of the 1930s may be seen as a short-lived experiment in the implementation of such an approach in a broad political arena.

Sweden was the country where, by all accounts, ideas corresponding to those of Rathenau, American progressivism, pragmatism, and New Deal politics saw their most determined fruition. In chapter 6, Elzinga, Jamison, and Mithander indicate how this was made possible by a combination of historical circumstances, including the success of the Swedish Social Democratic party leadership in capitalizing on certain cultural goods in the national heritage and turning these to their own advantage. The new "great power" project of a marginal group of young conservatives who appealed to nationalist sentiments was translated in such a way that the symbols of natural resources, technical innovation, and industrial prowess were linked to a social-reformist image of the "people's home" (*Folkhem*).

The concept of *Folkhem*, which had originated in the patriarchal culture of the traditional mining communities (the *bruk*), was given a new meaning that made it a vehicle for class conciliation on the basis of a historical contract of collaboration between labor and capital within a welfare state under Social Democratic governance. This made it possible to introduce regulative measures earlier and in certain respects to a larger degree than in other countries. It is no coincidence that it is in Sweden that one finds, in the 1980s, projects in which workers in the newspaper industry—typesetters, lithographers, graphic artists—joined with representatives from management and with university-based computer scientists to design new systems of computerized graphics that would give workers greater influence on the specifications for new technologies. Such experiments created a greater measure of democratic space for political deliberation than is normally the case. In the end, however, it became clear that as long as principles of private ownership over the means of production are not modified, democratic workplace input to and control over the shape of new technologies and how these affect workers' lives will still be limited (Winner 1995: 65–84).

From Convergence to Constructivist Theory

A leading theme in this book is that technological change is not something that occurs in the disembodied stimulus-response way in which it is often depicted in traditional accounts of economic and social history. Traditional narratives often depict an unproblematic process in which technology somehow develops and becomes increasingly important in the life of society while people constantly adapt to it in the workplace, in business, at schools, in the family, in their leisure time, and so on. This yields a linear picture of technological change as an autonomous process, with social impacts appearing as dependent and coming after the fact.

In the 1960s, convergence theory absorbed much of this perspective, not least when it came to discussions of technology transfer from developed to less developed countries. Convergence theory essentially boils down to three interrelated propositions: "(1) that technological development is essentially unilinear, that is, that there exists a relatively fixed path or sequence of technological innovation over which all industrial societies must travel in quest of their modernity; (2) that modern industrial technology tends to impose similar organizational constraints upon all societies, regardless of differences in culture, ideology or political institutions; and (3) that in the process of adapting to these universal constraints, human values and behavior patterns tend to converge in a technocratic ethos of 'amoral instrumentalism' " (Baum 1977: 315).

Of course, not all convergence thinkers would agree to all these propositions. Some might draw the line of the "technological imperative" at (1) or (2). In other words, we may distinguish three levels of alleged convergence: a technical, an organizational, and a behavioral level.

An early kind of convergence theory can be found in the Weberian idea of rationalization as a key to Western industrial societies. Weber saw the rationalization process everywhere, spreading through law, economy, accounting, technology, and so on and embodying an ethos that sought to subordinate everything to functional efficiency, measurement, and cost optimization. In Weber's theory of social change, rationalization replaced Marx's notion of class struggle as a driving force. As one might expect, Weber's theory was framed in the latter part of the nineteenth century, when the socialist movement had grown particularly strong and oligopolistic or organized capitalism had emerged as a formative trend.

In the course of the process identified by Weber, every society is supposed to move toward increasing bureaucratization, a tendency that goes hand in hand with the developmental trajectories of modern technology. Thus, like romantics and populists (but unlike Marx), Weber saw technology enabling new forms of domination, independent of the social system. In his *Wirtschaft und Gesellschaft* he wrote:

. . . the primary source of bureaucratic administration lies in the role of technical knowledge, which through development of modern technology and business methods in the production of goods, has become completely indispensable. In this regard, it makes no difference whether the economic system is organized on a capitalist or a socialist basis. Indeed, if in the latter case a comparable level of technical efficiency were to be achieved, it would mean a tremendous increase in the importance of professional bureaucrats. (Weber, 1913–14, cited in Bell 1971: 118)

Weber thus associated the growth and development of technology with a specific type of organizational structure and authority relationship. His may be called bureaucratic convergence theory, because of its emphasis on the bureaucratic element, which, he said, put an end to "politics" and replaces it with simple, machine-like administration. Weber saw this as a "universal" mode of development, leading to a unified code of values, behavior, and human relationships. Critics of technological development also often take this vision as their point of departure for articulating their worst fears and complaints.

The last few decades of scholarship in science and technology studies

(STS) have done much to correct the simplified pictures of technology assumed in various types of convergence theory, and it has centrally introduced the social dimension, with a constructivist twist. Bijker (1995b) has shown how STS as a genre developed in conscious opposition to contending theories of technology that it has sought to replace. Two of these earlier approaches are what might be called the *autonomous theory* and the *evolutionary theory* of technology.

The autonomous theory stresses the relative autonomy of technological development, as if it had an independent logic of its own. Models patterned on the autonomous theory are often deterministic, in the sense that they assume technological development to be unilinear, the same under all circumstances, and largely immune to social and cultural factors. This is different from the Weberian notion of convergence, since Weber, as Hård shows in his chapter, stressed the social roots of technology.

The evolutionary model allows for variation, recognizing mechanisms of internal technological adaptation and external adaptation in differing social environments. The technology follows trajectories that may vary in different economic, social, and cultural contexts, but the driving force is still taken to be technological change itself. An alternative form of this model stresses the pull of markets rather than the push of technology, in which case the external social and cultural factors are taken to signal information about suitable design to those working on the science and technology end of things. One way of making the evolutionary model more sophisticated is to graft on a recognition of the role of technological knowledge. In this case technological evolution is seen as a contingent process that incorporates both response to social environment and the ingenuity or intellectual capacities of engineers and scientists involved in the process.

Bijker's model is predicated on the premise that technology involves the social not only as an endogenous factor in an external environment but also in the very shaping of technical innovation from the ground up. Therefore, he speaks of the *social shaping of technology*, rejecting the notion of an autonomous, disembodied mode of development.

As many of the essays in this volume show, the idea of social shaping is not entirely new. Consider, for example, Lewis Mumford, whose view of technology is discussed by Jamison. Mumford postulated that certain socio-cultural conditions had to be at hand if a particular line of development was to take place. In other words, technological developments of particular kinds required, concomitantly, particular socio-cultural preconditions. In this sense Mumford stressed how technology does not follow its own momentum and is not a simple trial-and-error choice process, but rather is the result of a combination of socially shaping factors. This reflects a more general recognition that technology and culture interact, as do technology and politics.

Taking Mumford and Bijker seriously at the political level means that, if one wants to influence technological change, one must abandon the technology-out-of-control view that derives from the autonomous model. The depiction of technology as disembodied might also be called an "undersocialized" view of technology. Its adoption skews our perspective in favor of those established interests who stand to gain by an uncritical affirmation of prevailing technological structures and trajectories.

We must, however, also be wary of "oversocialization" (sometimes found in political science), in which technological change is treated as a dependent variable, with no cognizance taken of its contents or limits, let alone its dynamics. On this view we get a politics-in-command account, according to which technological trajectories may be steered at will.

Technology as Symbol and Image

A problem the authors of this volume have with Bijker's and other social constructivist models is that, for the most part, those models deal only with the micro level of technological change and leave the macro level outside their analysis. In order to deal adequately with the first crisis of modernity, we needed to address the role of the state, national politics, and technology assessment institutions. Our interest in this book is in the uptake of given technological systems in national economies, discourses, and cultures. In other words, we ask how technologies are deployed for various purposes, and how social and cultural factors are involved so as to win acceptance. We are dealing with what might be called the problem of the social and cultural appropriation of technologies, overall in public and private arenas within given countries, but also locally (as in the case of the introduction of mechanized grain unloaders in Rotterdam's harbor just after 1900).

In these cases we find that technologies are not taken up only *qua* hardware but also as perceptions in the minds of those who are involved or whose livelihoods are affected by the transformations they generate. New technologies have imagery and discourses created around them, and they are brought into the folklore of inhabitants, where they are linked to positive and negative symbols. In this process, existing symbols and cultural resources are invoked. Intellectuals and literary figures and other opinion leaders (the clergy in Dick van Lente's case of the Rotterdam grain unloader, female professionals in Catharina Landström's study of household technology) play important roles.

In other words, the introduction of new technologies involves not only new modes of organization of social relations but also a triggering of cultural nerves. Through this imagery linked to it in public discourse, be it in debates or through art and literature, a new technology is domesticated: it is actively made part of a repertoire of earlier and more familiar images that represent opportunities or threats. On the other hand, the symbolic linkage, once generated, may also operate in the other direction. That is, references to or debates around a new technology, a system of technologies, or even technology more generally may, in a discourse, come to symbolize many other things. Technology becomes a metaphor for everything good or everything bad in contemporary social trends. Thus, in our estimation, the best and perhaps only way to understand discursive patterns affiliated with technological change is historically and in a culturally comparative perspective.

Hård (1993) has gone some way to argue that what is needed are sociological perspectives that permit us to see technology more politically. At the level of associated imagery surrounding technological change, this means that we can look to metaphors and their use in a politics of representation within discursive patterns at different points in time. These will signify power relations and eventual contradictory interests vested in such change. Struggles over technological change may then be seen as means whereby groups seek to retain or rearrange social relations.

Traditional approaches in technology studies tend to assume a greater degree of homogeneity than what actually exists in societies. This also goes for Bijker's social constructivism, insofar as it interprets stabilization and the acceptance of technological trajectories in society as consensus processes involving closure. Instead, Hård maintains, technology should in many cases be regarded as the outcome of conflicting interests and ideas, and in this process culture is a distinct resource in the actual appropriation of new technologies to make them serve definite ends in a seamless web of techniques, ideologies, mentalities, traditions, and social configurations of which they constitute an intimate and even dynamic part. Formation of consensus is only one of several possible mechanisms of closure in debates or conflicts. Closure around particular technological trajectories or modes of design may also be the result of a domination of one social group or class over another. This is important to bear in mind.

It is this conflictual perspective that informs the present volume, where we are more concerned with the uptake and reception history of technologies than with their actual construction (though the socio-cultural preconditions for such construction remain of interest). Our focus thus foregrounds the interplay of technologies with various kinds of culturally loaded imagery and symbolism, because we see culture as a constituent ingredient in societal responses. In looking at the advent of technological change, it therefore becomes important to ask how such change is perceived by different groups in society, and how various metaphors and images are mobilized for or against supra-individual modes of modernization.

We have not attempted any systematic chronologies of events to compare crucial junctures or possible phases in comparative perspective over different countries, nor have we sought to reconstruct comprehensive patterns. Instead, we have thematized specific issues related to modernization, attempting to capture metaphors and images used to domesticate new technologies, and to explore contradictions or differences in perceptions among political, industrial, intellectual, and literary elites.

Intellectual and literary elites in particular are seen to play crucial roles in either overcoming or accentuating the sense of the alien in life with new technologies. A link between technological development and the shock of World War I is seen to be a feature common to many countries, although it is more traumatic in some than others. For example, in Sweden and the United States, which in different ways were more distant from the military battlefields, the conservative cultural response was weaker than in Germany, a loser nation.

What stands out is the fact that attitudes advocating social responsibility and rudimentary forms of regulation and technology assessment for the most part came too late—or too early, as in the case of Walther Rathenau. This continues to be true even today when we have elaborate institutions for such functions. The problem is that forces and institutional arrangements that affirm technological development and want to see it proceed more rapidly along established or mildly attenuated lines are separated from those forces and institutions that are supposed to regulate these same processes from the point of view of social costs and benefits for a majority of the population, or even minority groups whose livelihood and ways of life are placed in danger; the voices of the latter tend to be marginalized and their influence in *Realpolitik* tends to approach zero. This, then, remains one of the major dilemmas of organized modernism. Even under social democratic governance and the welfare state contract of class conciliation in the Swedish model there has been a division of labor and concomitant separation between affirmative technology promoting action and imagery on the one hand, and post hoc attempts at technology assessment and regulation on the other: the latter generally lacks the resources and clout allocated to the former.

In view of this state of affairs, where politics—both governmental and nongovernmental—has so little bearing on the technomorphic texture of modernization, it is not unnatural that frustration wells up, manifesting itself in anti-technology utterances and acts. At the cultural level, resisters have a wealth of traditional imagery and symbols in the heritage of their country on which they can draw in their opinion-building efforts. As Wagner notes, organizational change goes hand in hand with material change. I would add that this process also goes in tandem with the mobilization of cultural resources in complex discursive patterns in which technology becomes symbolic for a lot more than itself: freedom from drudgery, new social opportunities, savings in cost or time, a whole new world, or, contrariwise, regimentation, loss of values, alienation, and enslavement in a technified world.

Beyond Instrumentalism and Substantivism

Existential and moral issues emerge in the cultural appropriation of new technologies, both among proponents and among critics. It is no wonder, then, that technomorphic theories (such as Martin Heidegger's) that have been developed in critique of the modern condition continue to find so much resonance with each new generation. Heidegger is often interpreted as expressing humanistic worries about the domination of nature, which would place him in the same category as romantic critics of technology. However, as Hubert Dreyfus (1995) points out, Heidegger's critique runs much deeper than a mere critique of instrumentalist theories of technology as socially embedded functions that can be turned on or off, steered here or there. Dreyfus's is an ontological approach to technology, a technomorphic view, according to which the danger is not the destruction of nature or culture but restrictions on our way of thinking and being and a *leveling* of human understanding. Such a leveling can only be countered through a radical rethinking of the history of being in the West, which in turn can lay the groundwork for a philosophical reappropriation of technology whereby we are brought back to our authentic selves. Modern man, Heidegger writes, "must first and above all find his way back into the full breadth of the space proper to his essence. That essential space of man's essential being receives the dimension that unites it to something beyond itself. . . . that is the way in which the safekeeping of being in itself is given to belong to the essence of man as the one who is needed and used by being." (cited in Dreyfus 1995) Gaining a free relation to technology thus involves much more than reorienting relations of domination; it becomes a fundamental existential issue of how to distinguish and separate human destiny from human fate.

Heidegger's view has also been classified as a variant of a substantive theory of technology, as distinct from an instrumental theory. Instrumental theory is another name for theories of technology that depict science and technology as neutral tools that can be used for either good or evil.

Substantive theories, on the other hand, are ones that maintain science and technology to be so interwoven with human life, shaping it to such a degree, that we cannot just distance ourselves and turn technology's function around. Some large technological systems, for example military ones, can be used only for certain purposes (Winner 1980).

Andrew Feenberg has, on various occasions, argued for a third standpoint, which he calls a critical theory of technology. This is a theory that is supposed to steer clear of both euphoric affirmation and romantic resignation with respect to technological change. It recognizes the need to overcome cultural barriers and regulate development, but at the same time it renounces the illusion of state-sponsored civilizational change. The critical theory of technology derives from the tradition of the Frankfurt School. It agrees with instrumentalism in rejecting what Feenberg sees as the fatalism of substantive theories, whereunder he includes those of Ellul and Heidegger. On the other side, critical theory joins substantive theories in arguing that "the technical order is more than the sum of tools and in fact structures (or, in Heidegger's terms, 'enframes') the world in a more or less 'autonomous' fashion. In choosing our technology we become what we are, which in turn shapes our future choices. The act of choice is technologically embedded by instrumental theory. However, critical theory denies that 'modernity' is exemplified once and for all by the type of atomistic, authoritarian, consumer-oriented culture we enjoy in the West." (Feenberg 1991)

Critical theory, we are told, argues that technology is not a thing in the

ordinary sense of the term but an "ambivalent" process of development
suspended between different possibilities. Thus, it may also be called
the "ambivalence theory of technology."

Toward a Neo-Institutional View of Technology

Methodologically, rather than attempting to ground a macro-level
approach to the social shaping of technologies thesis in an aggregation
of micro-level studies, we find it more fruitful to start at the level of
institutions and incorporate the gains made by social constructionist
analysis. We bring in the cognitive dimension not as something appended
but as an integral part of what makes institutions what they are. It appears
that this is also the problem with which the "new institutionalism" has
been grappling in disciplines such as organizational studies (March and
Olsen 1989; Schimank 1995), where one finds a number of different
orientations, including some that make reference to cultural studies. For
our part, we seek to retrieve what has been learned about communicative
interaction—that is, the role of the cognitive order and sense-making
in the sustenance of such interaction within discursive patterns.

Neo-institutionalism points to the agency of actors in shaping the
environment they uphold by enacting norms that in turn shape their
identities and behavior, a process that involves the mobilization of cultural resources (Grenstad and Selle 1995). Culture is not a once-and-
for-all influence; it is an ongoing process, continuously constructed and
reconstructed during discursive interactions. In a recent book on the
interplay between two powerful American institutions, science and the
law, both of them seekers of truth, Jasanoff (1995) speaks of a "co-
construction" of ideas of truth and of justice in the context of legal
proceedings. More generally, we may speak of a co-production of scientific ideas and social order(s).

A second feature of neo-institutionalism is that it makes room for
individual entrepreneurs. These are seen as actors who can accrue
authority and power so as to become mega-actors, able to coordinate,
harmonize, or orchestrate changes in the societies in which they live,
often in interaction with other institutions and environments. Here the
ability to mobilize cultural elements in such a way that innovation
becomes incorporated and subsumed under patterns of the already
familiar is important.

Third, the assumption of cultural embeddedness of agency as well as technological change means that such change has a bearing on identity, cognitive codes, and values. Internalization of a norm system includes but is not equivalent to identification with the norms' sender or source. In identification lies an implication of drawing a boundary between us and them. In this way shared systems of meaning are produced and sustained, using language, folklore, metaphors, and symbols as well as what in some cases may be regarded as ritual ceremonies that communicate and constantly reenact key values, giving them credence in particular arenas of action.

Fourth, in this process whereby competent voices are tacitly or explicitly distinguished from those regarded as not competent, reputation and authority become important. The reputation and authority of leading speakers in a discursive frame have under certain conditions a tendency to accumulate and expand. In this sense, then, we can speak of the generation of credibility cycles, upon which turns acceptance or rejection of a given set of images and the technological change with which these are linked.

Credibility is, ideally, cumulative and distributive, as, for example, in communities that produce scientific knowledge. In their interplay with decision makers in a political arena, truth claims and images produced in scientific communities, when they have achieved a high degree of consensus (that is, when they are strongly stabilized), can be used to underwrite credibility and authority.

Science in relation to society can thus have a dual value, instrumentally in relationship to specific goals and by virtue of its symbolic value as expertise feeding into political communities and giving them greater clout. The greater the consensus and purity with which scientific knowledge claims are presented, the stronger their potential in the realm of power. Something of this tradeoff between purity and power (the image of truth speaking to power) lies at the base of the constant reproduction of the commonly held separation between science and politics, fact and value, or internalist and externalist factors in the growth of science and technology.

Finally, the foregoing processes, at a deeper level still, may be seen to generate and constantly recycle, reproduce, and maintain a moral order of fact and value claiming on each side of a debate or a conflict. Institutional or group boundaries in science are often constructed along disciplinary lines, which sharply define which individuals can make what claims. Similarly in politics and literature, interpersonal trust operates between individuals and groups so as to privilege certain speakers and disprivilege others in given contexts. Discourses on technology, then, are important not only to the acts of cultural appropriation but also for the moral order that holds societies and communities within them together, providing these with key elements of their identity and their self-understanding in a world of change.[1]

In the older forms of institutionalism cognition was taken to be channeled through values, norms, and attitudes. In neo-institutionalism, the latter are taken to be constantly in a process of social construction and reconstruction, upheld in tandem with the generation of affirmative and cautionary or rejectionist imagery surrounding change. Stabilization around particular arrangements, including the acceptance of new technologies *grosso modo*, then becomes routinized and self-legitimating around symbols that give change and action meaning (that is, a rationale or rationality).

As Di Maggio and Powell (1991: 11) put it: "Institutions do not only constrain options, they establish the very criteria by which people discover their preferences. In other words some of the most important sunk costs are cognitive." Cultural symbols are both media and sources of individual behavior relating to the affirmation or rejection of change. Cultural resources are thus central in processes of co-production of knowledge and social order. Likewise for technological change, for here too one finds a *co-production* of cognitive reflection and social order; this is what we look for in the discursive patterns relating to technological change in different countries, finding it natural that these patterns do not follow the same logic everywhere. Cultural goods that form part of the national heritage in various countries make a difference, and such

1. Daston (1995), in her work on the social history of rationality, facticity, and objectivity, refers to the "moral economy of science." This refers to mental states of collectives, or *Denkkollektiven* in Ludwik Fleck's meaning of the term. Moral economy does not refer to money, markets, labor, production, or distribution of material resources, but rather to an "organized system that displays certain regularities that are explicable but not always predictable in details." "A moral economy," Daston continues, "is a balanced system of emotional forces, with equilibrium points and constraints. Although it is a contingent, malleable thing of no necessity, a moral economy has a certain logic to its composition and operations. Not all conceivable combinations of affects and values are in fact possible. Much of the stability and integrity of a moral economy [in science] derives from its ties to activities, such as precision measurement or collaborative empiricism, which anchor and entrench but do not determine it." (ibid.: 4)

cultural goods, along with geopolitical positioning, level of technological change, and education, will differ and hence make a difference.[2]

Technological change at certain junctures, such as immediately after World War I, therefore may serve as a prism through which we can delineate more fundamental tensions in given societies. As certain physical, material, and social spaces are opened up thanks to new technologies and industrialization, others close. This opening and closing of spaces is not independent of culture but rather presupposes and involves it. Technological change, then, goes hand in hand with social reordering and cultural reinforcement or dissolution of the same. Technology is not a "bare" machine but also a "representation"; in the poetic, literary, or popular imagination the machine may be a strong metaphor invoked to cloud the oppressive nature of certain material realities and, failing this, to deflect negative responses to the level of only aesthetic protest. Technology may also be linked, metaphorically or symbolically, with great national projects that go far beyond the compass of the technical.

Thus, a leading theme, as has already been noted, is that technological change and its reception should be expected to vary from country to country, and that the substance of this process is contingent on diverse socio-cultural patterns and on history, including grand narratives in the national cultural heritage. We like to speak of this as the appropriation of technology, its domestication. In this process we see not only a social shaping of technology but also a *cultural appropriation*.

2. For a similar approach in the field of literature, addressing major macro-level issues of historical change and the role of culture in consolidating power relations and imperial hegemony, see the writings of Edward Said, for example *Culture and Imperialism* (1993).

3

German Regulation: The Integration of Modern Technology into National Culture

Mikael Hård

The First Crisis of Technology

In Thomas Mann's *Magic Mountain*, a young engineer, Hans Castorp, struggles to orient himself in a period of change. Mann (1924/1969: 155) writes: "Progress, up to now, had had to do, in Castorp's mind, with such things as the nineteenth-century development of cranes and lifting-tackle." The traditional engineering view of technology as the only truly progressive force, transforming nature for the material benefit of mankind, is no longer self-evident. Castorp has a hard time following the arguments of Ludovico Settembrini, an Italian man of letters. He cannot understand that values such as democracy and justice could also be seen as progressive. "All this made a confused impression on Hans Castorp. Herr Settembrini seemed to bring together in a single breath categories which in the young man's mind had heretofore been as the poles asunder—for example, technology and morals!" (ibid.: 155f.) That Mann highlights the transitory character of the "technology" concept in the 1920s is not the only interesting thing here. Perhaps even more exciting is that he attributes to Settembrini the idea that technological development may have positive moral consequences. It is the Italian *homo humanus* who puts forth the belief that new transportation and communication systems will bring about "the universal brotherhood of man." The humanist is the one who brings technical and cultural progress together—not the engineer!

Modern technology was much contested in Mann's Germany, often in a way that gave rise to surprising intellectual positions and coalitions. The struggle to find an appropriate place for "the machine" within the framework of German culture turned out to be a complicated one. Intellectuals tended to hesitate in the face of modern technology, and it is not surprising that Mann chose an Italian humanist as the representative of the most progressive and scientistic ideas. To hesitate is, however, not to oppose. As will be shown in this chapter, one of the most commonly held humanist positions in the German struggle was to call on the state to control and direct the development and use of technology. Many intellectuals, both on the right and on the left, wanted to organize technology, just as they wanted to bring order to what they saw as the chaotic economy of unfettered capitalism. In their attempts to organize modernity, they called for state intervention (see Wagner 1994 and chapter 9 in this volume).

The general crisis of classical, liberal modernity had already begun at the end of the nineteenth century. Modern technology, though, can be said to have gone through its first crisis during and shortly after the World War I. It became a central topic in the German discursive framework at this time. "The discussion about 'technology' became fashionable," one contemporary observer later wrote (Dessauer 1958: 27). Mustard gas and tanks proved to the critics that technology did not necessarily bring about a better world, and that it need not be an internationally unifying force. The postwar depression also proved to some commentators that capitalism had to be harnessed and directed toward common goals. Calls for planning, regulation, and control were widely heard—not least from Walther Rathenau, the multi-faceted AEG director,[1] government minister, and essayist. Although not a typical industrialist, Rathenau was just as concerned as everyone else about the economic problems that beset postwar Germany. Like most people, he was prone to blame the victors of the war for Germany's troubles.

It was not only former enemies that were victimized in the general debate; engineers and scientists were also criticized (Forman 1971; Herf 1984). Using the classical argument of technology's ethical neutrality, representatives of the engineering profession maintained that the engineers could not be made responsible for the cruelties of the war. In their struggle for renewed recognition, many well-educated engineers argued that their activities should be considered as "cultural" as those of artists and scientists (Dietz et al. 1996). In support of this argument they claimed that technology should be integrated into the national cultural identity, and that engineering education include components of classical *Bildung*. In other words, traditional frames of reference were mobilized to come to grips with modern technology.

1. AEG: Allgemeine Electricitäts-Gesellschaft.

The war did not, however, make all non-engineers antagonistic toward the machine system. It was generally accepted that technology was needed to reconstruct Germany after the war. To the influential social scientist Werner Sombart the war showed that technology had the potential to unify Germany and turn it into a great power. For the nationalist Sombart—as for the famous doomsday prophet Oswald Spengler—a prerequisite was that the German people discard foreign technology and develop their own nationally founded technology.

Nolan (1994: 9) points out in her analysis of the "incorporation" of American ideas into German society that few intellectuals were interested in importing foreign conceptions without modification. Even the most pronounced German advocates of Taylorism and Fordism argued that those originally American phenomena should be modified to fit German conditions. As Banta (1993: 14) puts it in her discussion of the narrative application of Taylor's and Ford's ideas within the United States, they were "appropriated by any ideology that found them useful." I argue that similar processes of intellectual integration and appropriation can be found in most German debates about modern technology. Even the x-ray engineer Friedrich Dessauer (1958: 62) saw it as his task—from the first decade of this century—to "incorporate technology into culture." He and other proponents of technical progress saw it, in my interpretation, as their aim to domesticate modern technology and assimilate it into traditional society without having to adapt to the demands of the mechanical spirit. Recourse was made to a traditional German discursive framework in which national *Kultur* was hailed and international *Zivilisation* was denounced. From Rathenau's "planned capitalism" in the 1910s to Sombart's "German socialism" in the 1930s, attempts were made to construct a "regulation discourse" and to control and organize the development of technology on a national basis (Dierkes et al. 1988). These ideas received strong support from the engineering community, and some of them were integrated into Nazi political practice (Herf 1984; Renneberg and Walker 1993).

Below I will present the positions of central German engineers and their classically educated opponents, as well as the views of Rathenau, Sombart, and Max Weber. Since it is impossible to cover the entire German "struggle about technology" within the limits of this chapter, I have chosen to treat these groups and individuals as representatives of distinct positions in the debate. There is, unfortunately, not enough space to treat either the socialists or the central Nazi thinkers at any length.

Aside from the fact that Rathenau, Sombart, and Weber were highly influential, those actors also initiated discourses that have since become the frameworks for contemporary academic and political discussions. The technology-culture problematic, which was so central to leading engineers in the 1910s and the 1920s, remains a topic in today's German-style philosophy of technology (Ropohl 1991). Similarly, the question of how to regulate technology is still dealt with in the 1990s in terms that are much the same as those that emerged after World War I (Dierkes et al. 1988). And the debate over technological versus social determinism continues to be central in social studies of technology from Sombart and Weber to Jürgen Habermas and Ulrich Beck (Rammert 1993).

Mandarins Domesticate Technology and Engineers Appropriate Culture

It is well known that German engineers constituted a self-conscious profession in the latter part of the nineteenth century (Gispen 1989). Hortleder (1974) has shown that many engineers actively strove for higher social status and fought to achieve high positions in the public administration. The engineers most active in this struggle for recognition were those who had gone through higher education at an institute of technology and had become Diplom-Ingenieuren (Ludwig 1974: 25).

The engineers did enjoy some success. Many polytechnic schools gained Hochschule status in the 1870s; these schools, in turn, were given the privilege of issuing doctoral degrees in 1899. The modern, science-based secondary schools were formally ranked on a par with the classics-oriented Gymnasien. This professional strategy turned out to be inade-quate, however. Not every engineer could become a civil servant or a famous business leader at a time when ever more engineers were being educated. The cry for higher status became hollow in the face of an emerging engineering "proletariat."[2] Since, furthermore, the social cri-tique of science and technology increased at the turn of the century (Hughes 1958), the engineers' situation became increasingly trouble-some. They were hard pressed to find other avenues for social recogni-tion. One such path entitled the engineers to adopt the philosophically idealist language common to those whom Ringer (1987) has called the German mandarins. This latter group consisted of the classically educated middle-class elite of government officials, university professors,

2. The total number of students at technical colleges increased from 3000 in 1885 to 11,000 in 1911 (Ringer 1987: 60).

secondary school teachers, medical doctors, and lawyers. Its members usually were politically conservative and held German culture and classical education in high esteem. They had gone through *Gymnasium* and had imbibed Fichte, Humboldt, and Hegel, thereby acquiring what Bourdieu (1991: chapter 2) calls a large amount of "cultural capital."

My main point in this section is that many engineers developed a new professional strategy after 1900, and that they did so in order to become accepted by the learned elite. Many engineers attempted to derive "symbolic profit" by adopting the mandarin discourse. Instead of primarily struggling for positions in the state administration and higher salaries, they chose a complementary strategy: to make modern technology palatable to the intellectual elite by adapting their vocabulary to the mandarin standard. The mandarins' power depended to a large extent on their ability to control and define what should be regarded as acceptable language (Ringer 1987: 21). Influential engineers attempted to foster the intellectual and cultural assimilation of modern technology by accepting the mandarin discursive framework in which the term *Kultur* played a central role. The engineers' attempts to give technology a cultural meaning can be seen as a "strategic turn on the conceptual level" (Ropohl 1991: 200). As Koenne (1979), Ludwig (1974), and Lenk (1982) have observed, the result was that some segments of the German controversy over modern technological developments came to be couched in the idealist language of the humanities. What these scholars have not discussed, however, is that the engineers' adoption of a mandarin vocabulary also affected the meanings of the terms being used. The machine entered the mind as an idealized entity, and thus were sown the first seeds of what would become *Technikphilosophie.* The discourse to which Ropohl himself belongs was framed at this time.

The concept of *Bildung,* which had its roots in the writings of Wilhelm von Humboldt, was central to the mandarin ideology (Bollenbeck 1994). Consisting of three interrelated elements (Ropohl 1991: 217), it emphasized individual personal development, prescribed Greek and Roman studies as the best training for the individual mind, and stressed that the spiritual rather than the material aspects of the world were most important. This idealism was, however, not contemplative and passive, but valued the constructive creativity of the human being (Ringer 1987: 28). Not only the intellectual but also the material outcomes of a truly creative existence could be called *Kultur*—and *Kultur* was something that the engineers were eager to get.

Throughout the centuries, strong mutual ties had evolved between the state and the academically trained intellectuals. The state had provided the intellectuals with certain privileges and had been able to mobilize administrative and rhetorical skills in return. At the turn of the century, this privileged position—the intellectuals' cultural capital—was threatened by a stronger currency. In the face of intense material, political, and socio-economic changes, the mandarins felt that their position as the carriers of spirit and culture was endangered (Ringer 1987: 12f., 50). The mandarins saw themselves being attacked by materialist engineers and scientists, inhuman industrialists, and (sometimes Jewish) businessmen. But even though the latter groups had acquired pecuniary strength, the mandarins still maintained that they should be entrusted with the preservation of classical values. Only they should be *Kulturträger.*

Toward the end of the nineteenth century the mandarins found their position ever more precarious, and they developed various responses to retain their socio-cultural advantage. This challenging situation forced them to review, rethink, and reformulate their ideology. Some mandarin "reformers," realizing that they now lived in "the era of technology," were willing to assimilate some components of modern life into the lower and middle levels of the educational system (Ringer 1987: 52f.). But more common reactions were to reject many of the constituents of modern society and to call for more state support. Most mandarins attacked rational technology and utilitarian science, as well as the mediocrity of democratic mass society, for being exponents of *Zivilisation.* To be civilized was, in the mandarin world view, to be superficially concerned only with the external world. The mandarins sharply contrasted this materialism with what they saw as the deep, inward-looking orientation of *Kultur.* They detested modern education, which implied the simple transfer of information and an orientation toward technological development and economic growth. Instead, they protected classical education as a means of cultivating self-development and personal growth. They wanted *Bildung* instead of *Ausbildung.* The right-wing writer Moeller van der Bruck summarized this position eloquently when he said that civilization belonged to the stomach and culture to the spirit (Stern 1961).

Toward the end of the nineteenth century, the mandarins began to defend culture for nationalist reasons too. They started to regard themselves as the carriers not only of inner culture but also of German culture. While civilization was denounced as a depraved Anglo-Saxon and French phenomenon, *Kultur* was praised as something genuinely

German. Thus, the term *Kultur* had no fixed meaning. Bollenbeck (1994: 29ff.) shows how, around the turn of the century, it became connected to all kinds of human activity, *Technik* included. By transforming its connotation from something spiritual and individual to something national and collective, the mandarins improved their chance of success-fully defending their position within a rapidly changing society. In the process, they revived a century-old tradition and modified it to fit a new era. Many engineering spokesmen began to do the same—but from another direction.

The cleavage in the German discourse between civilization and culture is well known. Less familiar, although by no means unknown, are the parts of the debate that were directly concerned with technology as a socio-cultural phenomenon (Herf 1984; van der Pot 1985; Ropohl 1991, chapter 10; Sieferle 1984). This discourse began to take shape in the 1900s, when the conservative mandarin writer Eduard von Mayer pub-lished a book with the title *Technik und Kultur* (1906) and when Friedrich Dessauer began to issue articles on this topic (e.g., Dessauer 1908). It was Dessauer who would later name this debate "the struggle about technology." This struggle did not primarily concern the positive or negative effects of certain innovations, as the nineteenth-century dis-course about "the machine" had. Rather, combatants from both camps maintained the traditional discursive framework, attempting to find a place for modern technology within the bounds of a classical and national culture. Thus, Ropohl's (1991: 198f.) assumption that all non-engineers subscribed to the technology-culture dichotomy is not correct, nor is his claim that this dichotomy belongs to a "particular 'German ideology' " or his suggestion that only non-Germans such as Lewis Mum-ford[3] have fully developed the idea of technology as a cultural product.

Mayer was opposed to much but not all of modern technology; how-ever, he did not uphold the traditional division between technology and culture. Through a concerted act of "intellectual appropriation" he redefined the concept of technology in order to make it possible to incorporate technologies he found to be acceptable into German culture while defining away and rejecting unacceptable technologies. Mayer did not embrace the classical standard definition of *Technik* as identical with material artifacts and machine systems. Instead, he counted all outcomes of genuinely human activity as *Technik* (Zschimmer 1937: 27–37). Tools, factories, books, and even social systems now belonged to technology—

3. See Jamison, chapter 4 in this volume.

granted that they were products of true personality. Mayer followed the ideal of individual *Bildung*, and argued that the best technology would enable everyone to develop his or her own abilities. His view was, however, not individualist in the liberal sense. It was, rather, inspired by communitarian ideals, and Mayer demanded that technology contribute to social and cultural unification. Modern technology, however, did not aim in any of these directions. Instead, this technology created an atomized and inhuman world in which there was no room for personalities to develop.[4] Mayer criticized the destructive and demoralizing power of the modern factory system. He blamed the modern "technical spirit" for causing "alienation" and for constructing a society in which uniqueness and individual needs were run over by standardized solutions for the masses (von Mayer 1906, quoted in Dessauer 1958: 122).

For outspoken engineers such as Dessauer, such attacks had to be opposed. Because the debate was no longer about directly political, legal, and administrative issues, such as government openings or university degrees, a more intellectually and cognitively oriented strategy was called for. In his subsequent responses to Mayer and others, Dessauer went out of his way to make his arguments and his language fit into the discursive framework of the mandarins—to meet them on their own turf, so to speak. In the process, however, he (like Mayer) also tried to affect and modify the content of this framework. When he was debating the relationship between technology and culture with a cultural historian, one part of his strategy was to expand the connotation of *Technik*. He wanted to give the term a spiritual dimension. Paraphrasing modern discourses, one might say that Dessauer (like Mayer) did not want to limit the term *Technik* to hardware. He accepted the mandarin idea that *Geist* was an integral part of true *Kultur*, but he claimed that technology also contained *Geist*. With this move, technology was given both a material and a spiritual side.[5]

The expansion of the term *Technik* was more than wordplay. Since it enabled the engineers to claim that engineering was just as spiritual an

4. More than 30 years later, Eberhard Zschimmer launched a full-scale attack on Mayer. Zschimmer (1937: 37) sees Mayer as a representative of the so-called cultivated (*gebildete*) thinkers who had departed so far from reality that they had become deformed (*verbildete*).

5. Note that the discussion is here about *Technik* and not about *Technologie*. The latter is an epistemological and educational term that, since Johann Beckmann, has had to do with knowledge about *Technik* (Müller and Troitzsch 1989).

activity as studying history or reading the Bible, it may be seen as an attempt to make technology acceptable to the learned elite. In the following decades, several other attempts in the same direction were made (see, e.g., Coudenhove-Kalergi 1922; Stodola 1932; Zschimmer 1914).

The reactions of German intellectuals to World War I were divided. Sombart and Spengler were not alone in finding that the war disclosed the hidden potential of German culture and the German nation. The young right-wing writer Ernst Jünger regarded the war as a purge that would ultimately strengthen Germany. Jünger was not a traditional mandarin; he was an independent writer and a hard-core representative of what Peukert (1987: 26ff.) has called the "frontier generation." He volunteered to fight in the war, and the experience proved to him that modern technology was one of the principal means of selecting the strong members of society (Herf 1984: 70ff.). In a futuristic manner, Jünger wrote: "Ours is the first generation to begin to reconcile itself with the machine and to see in it not only the useful but the beautiful as well."[6]

Other intellectuals were not so positive. During the war years the mandarin theologian and medical doctor Albert Schweitzer formulated a radically pacifist ethic that included a fundamental critique of modern technology. Schweitzer's view was not, however, that the war was the root of the present crisis. Rather, it was that both the war and modern technology were the outcomes of imbalanced cultural developments. "The destiny of our culture," he wrote, "is that its material side develops much more strongly than its spiritual side." (Schweitzer 1923–24, II: 2) This imbalance was manifest both in the cruelties of the trenches and in the prison-like working conditions of the modern factory. Schweitzer denounced modern technology for treating human and other living beings like things, in an instrumental manner. He called for a technology based on ethical considerations. In Weberian terms, one could say that Schweitzer wished to substitute a value-rational technology for the dominant purposive-rational technology. Schweitzer (1923–24, I: 21) wanted a technology that would relieve both "individuals and collectives" of the struggle for immediate survival. Following the *Bildung* tradition, he asked for a technology that would make people free to cultivate their own abilities.

6. Quoted from Herf (1984: 70); the original passage is in Jünger 1929.

Like Mayer and several other mandarins, Schweitzer redefined the concept of *Kultur* to meet a new situation. Individual freedom had always been an important ingredient in the mandarin frame of reference, but Schweitzer now reformulated this frame to fit his pacifist and non-elitist ideology. His critique of elitism is noteworthy, since most mandarins were elitists. He defined *Kultur* as an ethical position that guarantees all living beings freedom and due respect. A truly cultural technology would liberate humans and animals, rather than exploit and threaten them. Schweitzer's *Kulturphilosophie*, written during the war but not published until several years later, can be read as a full-scale attack on the war effort in general and on military technology in particular (see Schweitzer 1923–24, II: 270). But *Kulturphilosophie* can also be seen as an attempt to gain cultural control over technological development and to lead it in new directions. Schweitzer did not reject technology, but he wanted to appropriate it for new, more spiritual purposes. He wished, so to speak, to domesticate it in order to liberate all living beings.

The mandarins were not the only ones to be scared by the enslaving and brutal character of military technology and the factory system. Even though most engineers probably were not directly afraid of the present technological trajectories per se, several of them were definitely concerned about the strong critique that the negative consequences gave rise to. If the painful impacts of technology became more pronounced in the debate, and if the critique consequently grew stronger, then their profession and cause could be in danger. One writer who realized this[7] was Manfred Schröter, the son of an engineering professor, who in the 1910s developed a philosophy that treated technology as a cultural product. In contrast to the traditional mandarins, Schröter (1920: 49) did not regard philosophy, art, and religion as the only constituents of *Kultur*; he also assigned science and technology to "the totality of culture." In his redefinition of culture he picked an aspect that had always been inherent in the humanistic tradition: *Kultur* and *Bildung* as creative rather than contemplative activities. Schröter defined culture as all actions that bring order and meaning to human life (ibid.: 1). Religion and philosophy had been means of creating such order, but in modern society their tasks were increasingly fulfilled by science and technology. Schröter consequently did not limit *Technik* to tools and machines; he made it encompass all "creative, productive work" aimed at giving life meaning by material means (ibid.: 56f.). In other words, he made use

7. See Herf 1984: 164ff.

of the mandarin discursive framework to appropriate technology into a redefined concept of culture.

Schröter was not altogether positive toward the developments in his own time. Interestingly, his criticism of the enslaving character of much modern technology, which comes out most clearly in his *Philosophie der Technik* (1934), echoed Schweitzer's views. If technology is not governed by spirit, it may take a dangerous turn. The "functionalist" pro-engineering writer and the humanist mandarin could indeed find some common ground.

Spengler's influence on the public debate was greater than that of anyone else at this time. His *Decline of the West*, the first volume of which was first issued in 1918 and the second in 1922, triggered an intense debate both inside and outside Germany. In defense of Spengler, Manfred Schröter (1922) even wrote a book that summarized this "struggle." Spengler was an independent writer rather than an established mandarin, and many scholars criticized his historiographic approach. But several of his ideas and conclusions did nevertheless fall on fertile ground, especially among orthodox mandarins (Ringer 1987: 204). Spengler adhered well to the mandarin discursive framework. He picked up and reformulated the Goethean tradition, and he was heavily influenced by Nietzsche. He rehearsed the familiar dichotomy between *Kultur* and *Zivilisation*, and he blamed the corrupt Western civilization for most of the depravities of modern society. Like Rathenau, he detested *laissez faire* liberalism and argued for an orderly and controlled economy. Like Sombart, he wanted a new kind of socialism based on German virtues.

Spengler devoted the very last chapter of his well-known book to "the machine," and he would later return to this topic in a book entitled *Man and the Machine* (1931). He defined technology very broadly as every ability to act, including even the ability of an animal to run (Spengler 1991: 1183). Technology was not in itself a problem, only capitalist technology was. When the "idea of the machine" emerged in the Renaissance, it had the potential to transform the world into something great, something that incorporated the truly creative soul of mankind (ibid.: 1187). Owing to the devastating influence of merchants and business leaders, it had subsequently run amok, however. The inhabitants of the West had become technology's slaves rather than its rulers. And the otherwise creative engineer, "the enlightened priest of the machine" (ibid.: 1191), had become its victim.

One reason that Schröter felt a need to defend Spengler was that *The Decline of the West* was commonly read as anti-technological through and through. Phrases like "the machine belongs to the devil" were broadly misunderstood, both by Spengler's contemporaries and by later scholars (see e.g. Hortleder 1974: 86ff.). As Herf (1984, chapter 3) has pointed out, it was generally not observed that Spengler tended to locate the worst aspects of modern technology in the most overtly capitalist countries in the world, and that his book included a plea for a technology not governed by Mammon. Spengler's traditional heroes were the farmer who cultivated the land and the blacksmith whose goods incorporated the soul of their maker. Their activities were extracting (*erzeugend*) and refining (*verarbeitend*), whereas the "exploiting" merchant only transmitted (*vermitteln*) goods (Spengler 1991: 1158ff.). Spengler's ideal was a technology controlled by the really productive people, such as craft workers and engineers (the heirs of the blacksmith). He wanted a technology that grew out of concrete experience and from a creative soul rather than an exploiting mind. Only such a technology could be truly cultural. For the German people, this implied a need to reject the technology of France, Great Britain, and the United States and create their own technology—one based on German *Kultur*.

In line with the main perspectives of this book, it is possible to interpret Spengler's work on two levels. First, Spengler tried to domesticate and tame modern technology by incorporating it into German culture. Second, he attempted to bring order to capitalist modernity by creating a new kind of communitarian order (Wagner 1994).

Spengler and many mandarins were by no means alone in expressing a fear of foreign technology and arguing in favor of a nationally oriented technology. Even some influential engineers did so. The *Diplom-Ingenieur* Carl Weihe asked his colleagues not to sell out to international technology and warned his countrymen of the dangers of "*Amerikanismus* and all its detrimental effects on the people's soul"—whereby he meant the soul of the German people (Weihe 1919: 86). This fear of what was considered American ideals and technology started to become quite widespread after World War I (Hermand and Trommler 1988: 49–58). Weihe (1919: 86) implied that modern technology could indeed be a threat if it were not founded on "socio-ethical thoughts." He urged the men of technology and industry to do anything in their power to develop a humane technology and to propagate knowledge about the benefits of technology to the citizens.

A couple of years later, Weihe became editor of the *Zeitschrift des Verbandes deutscher Diplom-Ingenieure*, the name of which he changed to the less boring *Technik und Kultur* in 1922 (Herf 1984: 171–178). Weihe used this journal very effectively as a platform for an intensified fight for public acceptance of engineering as a cultural endeavor, becoming one of the most influential and prolific spokesmen for the engineering corps in the interwar period. His attempts to make technology "the darling of the people" had begun just after the war as a reaction to the anti-technical sentiments at that time (Weihe 1919: 87). He agreed with many critics that "guns, cannon, airplanes, men-of-war, and submarines" were terrifying, but argued that they could not be taken to be representative for technical products as a whole (Weihe 1918: 330). On the contrary, he argued, history showed that technology incorporates cultural values and contributes to spiritual liberation, since it freed people from some of the drudgery of hard labor (ibid.: 331).

Weihe supported the engineers' cause by making modern technology more German and more ethically acceptable, and he adopted the mandarin discursive framework by analyzing technology in terms of *Kultur* and *Geist*.

Although many engineers began to adjust themselves to the mandarin discursive framework after 1905, their reasons for engaging in public discourse were the same as they had been at the end of the nineteenth century. Professional interests still guided their struggle. Dessauer (1958: 22) declared explicitly that in his activities as a writer and speaker he had always aimed to enhance the social status of the engineering profession.[8]

Bourdieu (1991, chapter 7) points out that there are no neutral means of communication. Whereas those who control the material production in a society have economic power, those who control the formulation of ideologies and other world views have symbolic power. Symbolic power will remain in force as long as the subordinated groups can be made to believe in the ideologies produced by the ruling classes. Here is the problem that faced the mandarins at the turn of the century. Their ideological monopoly was challenged, and only some modernist reformers understood that it could be self-defeating to cling stubbornly to all the classical ideals. Instead of rejecting modernity outright, Mayer and Schweitzer chose to make some of its aspects fit into a redefined version of *Kultur*. One part of this strategy was to domesticate and tame modern technology by bringing out the constituents that were specifically German or particularly liberating. The result of this strategy was the formulation of a discourse that is still with us today as *Technikphilosophie*.

8. I follow the suggestion of Childers (1990) that the German engineers acted as a profession in the modern, Anglo-Saxon sense, although they adopted a premodern vocabulary in that they characterized themselves as a *Stand* (estate).

Community, Order, and Planning: Rathenau's "New Society," "New State," and "New Economy"

Walther Rathenau symbolized the German struggle for postwar recovery and rectification more than anyone else. He fought to survive in a situation of scarcity and hardship as director of the large electrotechnical company AEG, as the man responsible for organizing the national supply of natural resources during World War I, and as Minister for National Reconstruction in a postwar government. His activities as administrator, businessman, and politician are fairly well known and will not be the focus here. Instead, attention will be directed at the visions of a future society he sketched after 1915, rather categorically dismissed by Barnouw (1988: 48, 66) as "flawed" and "utopist in the old static meaning of the concept."

Rathenau tried to come to grips with wartime chaos and postwar depression by means of rational planning toward common goals. He envisioned, from his position at the very top of German society, a strong state and a nation where the power of special-interest groups had disappeared. Although he invested his hope in the youth (Rathenau 1918a), his heart—like those of many others in the "Wilhelminian generation"—was still with the ideals of Bismarck and the *Kaiserreich* (Peukert 1987: 26).

Rathenau formulated his vision in the numerous books, pamphlets, and articles he wrote between 1915 and 1922. Titles like "The New Society," "The New State," and "The New Economy" reveal that his goal was a new start (Rathenau 1918b, 1919c, 1922). He had already made a name for himself as an essayist in the first decade of the century, but it is clear that the war affected him profoundly. Shortly before 1914 he had written two philosophical critiques of the contemporary *Zeitgeist*, but his writings during and after the war highlighted economic and political matters. It is true that he began his 1917 book *In Days to Come* by claiming that material phenomena are of no inherent value and that their only purpose is "to foster the development of the soul and its realm," but in his subsequent analyses there was remarkably little talk about the soul and the spirit. (See, e.g., Rathenau 1917a: 5.) Instead of delivering an idealist "critique of our age" or a Bergsonian analysis of "the mechanics of spirit," as he had done before the war (Rathenau 1912, 1913), he now discussed how Germany should be organized to meet the economic, political, and social challenges that followed in the wake of the war. He wanted to guide Germany through a disastrous situation in which its national finances were in ruins, its constitution was being questioned, and its population had been substantially reduced (Rathenau 1918d; 1918e: 57). And in such a situation Rathenau found that a pragmatically economic discussion was most pertinent.

Rathenau was a prominent figure in German high society. This does not mean, however, that his ideas always squared with those of the political, military, and business elite. Barnouw (1988) and Hellige (1990) contend that Rathenau was an outsider who embraced some of the views of his peers and rejected others. He was a civilian, but he became responsible for the mobilization of natural resources for the war effort; he was a capitalist (for instance, a member of numerous boards of directors), but he argued for more state power; he was a Jew but a devout German nationalist; he was an engineer, but he warned of the dangers of mechanization.

Against this background it is not surprising to see that one of Rathenau's main targets in his postwar writings was the liberal economy organized according to *laissez faire* principles. The "uncontrolled struggle" between German firms led only to chaos and to mismanagement of resources (Rathenau 1918e: 67). It was appalling that "hundreds of thousands of men travel about on railroads only to carry out competitive struggles between various trading companies" (ibid.: 75). Competition was generally disastrous to the wealth of a nation. It was morally irresponsible that each individual was striving for his or her own benefit while Germany is bleeding to death. Rathenau's critique of an overly liberal economic system thus had a distinctly nationalist bias. He disliked struggles within Germany, but he accepted economic competition among nations.

Rathenau was in favor of most kinds of domestic monopolies and had argued for organizing the German economy in cartel-like bodies during the war. He found it quite correct that "the owner of a railroad, a waterworks, a harbor receives monopoly rights directly from the state or the municipal authorities" (Rathenau 1917a: 107). Some people might believe that such monopolies—along with patent rights and cartels—were unjust and inefficient, but Rathenau argued that the opposite was indeed the case. At best, competition was a zero-sum game. The "monopoly of technological advance" was particularly important for the creation of common wealth (ibid.: 108). Competing companies might dislike being unable to share the fruits of new inventions, but patent rights enabled their holders to utilize innovations in a way that was economically rational in relation to society as a whole.

Rathenau realized that a prerequisite for his economic system was either that it be imbued with "a moral sense of responsibility" (1918e: 67) or that it be controlled in detail. Monopolies might otherwise be misused, and people would work only for their own benefit. Each individual and firm had to feel obliged to work for the common good of the nation. Indeed, only the sick and the elderly ought to be allowed to remain idle (1917a: 131). Laziness, speculation, and shortsighted egotism ought to be banned.

Despite these restrictions, Rathenau never questioned "the law of capital" (1917a: 70). He sometimes called himself a socialist, but he did not see Soviet communism as an alternative to liberalism. He did not want to socialize the means of production, but he wished to prevent private companies from acting against the interests of the country as a whole. He did not want to do away with the right of private ownership, but he disliked inherited wealth and privileges (1918c: 38). He did not want to strangle world trade, but he argued in favor of substantially increased customs duties. His vision included a system with private but "not unfettered" business (1918e: 67).

Rathenau was well aware that materials, energy, and labor were limited in supply, and he regarded the utilization of these resources as a common, societal concern (1917a: 247). Every citizen had the right to a certain share of natural resources and the fruits of industry. For reasons of both economy and equality, he thus wanted to reduce the consumption of unnecessary items. He stigmatized the vanity of the upper classes and claimed that their greed contributed to the increasing poverty of the lower classes: "... all those years of work, which are required to produce an expensive piece of embroidery or a magnificent tapestry, have forever been withheld from the clothing of the poor" (1917a: 78). While some people immersed themselves in luxury, others did not have enough to eat. This was especially disastrous in the case of imported goods, since resources were not only taken away from the poor but also taken out of the country. The importation of luxury items had made Germany poor and should be avoided (1918d: 26; 1918e: 55).

Unfortunately, indulgence in luxury was not limited to the most well-off segments of society. Rathenau claimed that soulless hedonism was spreading. Many people were ashamed of yielding to "conspicuous consumption" (Thorstein Veblen's term), but they nevertheless tried "to cover this shame by weaving some feeling into the miserable technology of today's life and its products" (Rathenau 1917a: 8). People demanded products with a "colored surface" in order not to be criticized for being overly materialist (ibid.: 9). In other words, superfluous design had become a means of making up for unlimited consumption. Reducing the demands for new goods seemed to be out of the question.

The main carrier of Rathenau's almost revolutionary vision was to be the German state. He was unusually poetic when describing this state:

> . . . we see beams of light touching the picture—beams from the sun of the state. We do not have the socialization of the economy very much at heart, nor a mania for state intervention, where such is not needed. But the feeling grows ever stronger that we are not only economically responsible for ourselves but also for each other, and that we and the state are mutually responsible. A more thorough feeling of community between the state and industry is not frightening, granted that the state is able to liberate itself from one-sided and bureaucratic methods . . . and grow to the highest and most genuine organ of the common will and spirit. (1918d: 28)

Only the state, in Rathenau's view, could guarantee that common needs would match private interests. The state would regulate economic matters while still allowing freedom of thought, speech, and religion. The new state would make sure that production and distribution were efficiently executed and excessive consumption avoided. It would, for instance, introduce high import duties, increase taxes on luxury items, and give each citizen a minimal share of the common wealth (1917a: 103). Sales taxes ought to be increased and income taxes reduced. Rathenau's "new state" included both paternalist and "neo-mercantilist" (ibid.: 251) elements.

Rathenau regarded as the great enemy of his state the massive number of special-interest groups that fought only for their own benefit. Particularism, he suggested, was dysfunctional to the organic unity of any people and to the collective wealth of any nation. However, this should not be taken to imply that Rathenau was opposed to all kinds of group influence. He valued *Sachlichkeit* (impartial matter-of-factness) highly, and he wanted each professional group to be able to affect the organization of its area of expertise (1922: 29). For instance, he found it quite natural that professors and other teachers have a say in educational matters, but he argued that they ought to do so as experts rather than as special-interest groups (ibid.: 39). Since Rathenau regarded *Sachlichkeit* as an integral part of "the true German spirit," he was convinced that it would be possible to separate expert knowledge from subjective interests in Germany (ibid.: 9). His state would not only be paternalistically organic;

it would also be technocratically corporate: "We will erect a state which is more impartial, organic, righteous, free, and efficient than any other state. . . . That is the New State." (ibid.: 71)

Rathenau's state was to rest on the shoulders of the German people. The problem for Rathenau was that they were socially disunited and had become "mechanized in spirit and ways of living" (1918e: 72; 1922: 16). It thus became necessary to unite the people. Although a Jew, Rathenau never ceased to emphasize the unity of the German people— Teutons or Jews. Like several conservative Swedish critics in this period (see Elzinga et al., chapter 6 in the present volume), Rathenau (1922: 10) talked about and wanted to restore "the great time" in the history of his country.

Rathenau was well aware that the material basis of the new society could only be based on industry and agriculture. Since industrial production was generally wastefully organized, a "science of industry" was required (1918e: 76). As long as most firms were able to sell whatever they produced at a great profit, industrialists had few incentives to save resources. Carried by well-educated engineers, a "scientific technology" had made its way into only a few branches of industry—the power industry, where Rathenau himself was most active, being one of them (ibid.: 78). The passive attitude of most firms might have been understandable if the economy had been booming, but it could not be accepted when "all losses and waste [were] of common concern" (ibid.). The scarcity of resources that had become obvious during the war had made it immoral not to modernize production.

Numerous interpretations of Rathenau's writings and deeds exist. Mader (1974) has called him a "functionary of finance capital." Hughes (1990) has suggested that Rathenau was a "system builder" who not only erected technical systems but also tried to organize firms and whole branches of industry in a systemic fashion. Paradigmatic in Hughes's interpretation is Rathenau's attempt to coordinate the German electro-technical industry by means of state intervention; Rathenau tried to create a national monopoly in this sector for the sake of optimal efficiency. Hughes's explanation of Rathenau's centralization policy has subsequently been criticized by Hellige (1990), who claims that Rathenau's attempts to regulate industry had to do with the negative experience he had had in his youth of unrestricted competition in the chemical industry. Rathenau was, in Hellige's analysis, an "organizer of capitalism" rather than a "system builder"; he was closer to the US engineer and socialist Charles Steinmetz and his "corporate collectivism" than to Thomas Edison and his system-building strategy.

Rathenau's project could also be interpreted as fitting into what Galbraith (1967) has called "organized capitalism," or as attempting to introduce a "visible hand" to the market (Chandler 1977). However, I believe that all such interpretations are somewhat skewed in that they take Rathenau's business experience as their point of departure. Rathenau's perspective did not involve only a business policy; his vision encompassed not only the world of production but society as a whole. It is certainly no coincidence that his writings and his wartime planning policy inspired Lenin and other Bolsheviks in their plans for the Soviet Union (Maier 1987: 43). We have to look closer at the political dimensions of his vision. I would argue that Wagner's 1994 analysis of the demise of "liberal utopia" at the turn of the century is a more appropriate basis for an interpretation of Rathenau's work. Wagner (ibid., chapters 5 and 6) suggests that classical modernity in general and classical liberalism in particular were replaced by "organized modernity" and various statist projects during the interwar period. The visionary ideas of the French Revolution had failed. Free enterprise, true democracy, and impartial science had proved impractical. Instead, various actors tried to curb business, politics, and science. The liberal idea of free human beings gave way to collective notions such as class and *Volk* (Stråth 1990). Of course, Soviet Communism and German National Socialism came to epitomize these trends, but less extreme attempts to control modernity also developed. In Sweden the *folkhem* vision was formulated by both right-wing intellectuals and leading Social Democrats (see chapter 6 below). Americans got the New Deal. Wagner (1994: 67) writes that "all these proposals were responses to the perceived instabilities of the post-liberal regimes" and that "they were all based on the definition of a, mostly national, collective and on the mobilization of the members of such a collective under the leadership of the state."

I conclude that Rathenau's ideas and activities during and after World War I constituted a concerted attempt to regulate the modern project without rejecting it. Rathenau strove to bring industries and individuals onto one path, where their efforts could be collectively organized and directed toward common goals. He wished to substitute discipline for liberty and community for ego. His policy was aimed at the assimilation of modern technology into familiar structures rather than at the adaptation of these structures to the demands of technology.

The war opened up in Rathenau's mind the door to a world without egotism. Its cruelties notwithstanding, it had some positive consequences: ". . . the world fire has borne fruits that—in the face of cold egotism and indifference—otherwise might have taken centuries to grow ripe" (Rathenau 1918e: 71). As early as 1915, Rathenau (1917a: 77) observed, the resistance of the liberals toward state intervention had begun to weaken as a result of the war. He hoped that the war would ultimately convince each individual and each firm of the necessity of state coordination, national cooperation, and hard labor toward common goals. Only countries imbued with "planned order, conscious organization, scientific thoroughness, and a loyal feeling of responsibility" (1918e: 67) would survive in the long run. The rest, "the dross," would disappear (ibid.: 72).

Rathenau thus attributed to the World War the role of a Great Divide in history, and he regarded it as an event with ultimately positive results. Unlike Schulin (1990: 55), I cannot see that Rathenau was "unsure" whether the war would contribute to the "revolutionary changes of the modern world" he had been waiting for. Rathenau was quite certain that the war would have drastic consequences. The war had led to the introduction of new production methods and the construction of new plants in the chemical, metallurgical, and shipbuilding industries (Rathenau 1918e: 50). But, more important, it brought forth a new society. A new global economy, a new industry, and a new Germany would rise out of the ashes. "Property, consumption, and demands" were "no private issues" in this world (Rathenau 1917a: 75). Because of the scarcity of natural resources, the German people would have to learn to economize. This did not mean that Germany should turn its back on modern industry, but rather that industrial production had to be made more efficient. Better planning and improved organization were needed. At the center of Rathenau's vision were the concentration of production to fewer but larger units, the standardization of products in order to limit unnecessary large varieties, the mechanization of production, and the application of scientific knowledge (Rathenau 1918d: 27f.). Here, it was not only the state that had important tasks. Firms within the same branch of business ought to cooperate rather than take part in meaningless competition. They ought to carry out programs of structural rationalization in order that production be made optimally efficient. Through such programs, a "scientifically organized group-division of labor" between firms should be arranged (Rathenau 1918e: 80). Presumably with the IG-Farben cartel in mind, Rathenau (ibid.:

111ff.) praised the chemical industry for having gotten the farthest in this regard.

The war gave Rathenau good reasons to argue in favor of various rationalization measures. Although he did not himself employ the term "rationalization" when discussing ways to make the economy more efficient, many of his ideas squared well with those of the emergent "rationalization movement" (De Geer 1978). At the core of this movement were Frederick Taylor's "scientific management," Henry Ford's assembly line, and various forms of corporate restructuring (e.g., cartel formation). All these measures were beneficial to industry in general and to large companies in particular—not least to Rathenau's AEG. My interpretation is that the war made it possible for Rathenau to present these special interests in the guise of common, general interests. Such a strategy was quite persuasive in the German discursive framework once its American roots were played down. The largest part of the upper and middle classes in Germany—not only the mandarins—found the state to be both objective and fair. If anyone had the ability to make sure that modern industry would be organized and tamed in an acceptable way, it would be the state (Wagner 1994, chapter 4).

Since the new economy would be nationally organized, it would be possible once again for national characteristics to affect technological developments. Rathenau (1918d: 20) lamented that mass production had led to an increasingly uniform technology: "Goods flooded the Earth. The old, beautiful differences that used to encounter anyone visiting distant cities and countries have been lost in our age. No longer is the traveler able to experience and appreciate the fruits of new soils, art, and labor."

Rathenau concluded that, if production were to become less international, local styles might reemerge, and it would be easier to resist the American influence on technology. German technology would reflect the German mentality to a larger degree and would thus be more familiar to the German people.

Thus, Rathenau tried to come to grips with resource scarcity and uncontrolled competition by means of a strong state, and to make modern production technology and capitalist exploitation acceptable through the use of nationalistic argumentation. Bismarck had seen a strong state as a tool for the unification of the German people 50 years earlier, and Hindenburg had used the machinery of the state to provide the armed forces with material resources during the war. Rathenau, however, regarded the state as a means of making modern industry attractive. Under the wings of the German state, modern technology, imbued by a nationalist spirit, would be domesticated.

A central thesis of this volume is that various events in the period 1900–1939 challenged Western intellectuals in profound ways, and that they met these challenges by making recourse to familiar but reformulated concepts and themes. In the German case, World War I was the watershed. A comparison of Rathenau's postwar writings with two essays he had published just before the outbreak of the war supports the thesis.[9]

The main target of Rathenau's emotionally engaged *Zur Kritik der Zeit* (1912) and *Zur Mechanik des Geistes* (1913) was the soulless, mechanical character of modern life. It had become necessary to make the production of basic goods more efficient, but "the contemporary technical habitus" has spread to other areas of human activity (Rathenau 1918f: 4). The "soul" had become increasingly marginalized in a world ruled by "business, diplomacy, technology, and transportation" (Rathenau 1918g: 37). Rathenau argued by means of classical dichotomies from *Lebensphilosophie* and claimed that the world rested on rationalism and intellectualism rather than idealism and intuition. It was driven by spirit rather than soul, and its central values were efficiency and utility rather than quality and truth. It was a mechanical world.

The root of the tendency toward a mechanized society was the rapid population growth that had taken place in recent centuries:

It is important to notice in passing that technology or transportation are not the cause of mechanization and our contemporary forms of life. Rather, increasing population density has driven mechanization forward, and the mechanization process has given birth to new needs and created new aids. It would be wrong to claim that the railroad has brought to life large traffic flows, or that the rifle has led to mass war. (Rathenau 1918f: 30)

Rathenau's analysis was thus directed against all forms of technological determinism, even though it had a historically materialist bent. Organizational and technical innovations intended to ensure a continuously increasing degree of efficiency had been developed in response to population growth. Stimulated by competition for limited resources, the capitalist "law of production" had proved especially efficient and had led to ever-higher speed and turnover (ibid.: 39). In addition to large amounts of identical products of poor quality, capitalism had brought

9. Schulin (1990: 55) writes that the war caused Rathenau to "question his previous ideas, plans, and focuses."

specialization and abstract thinking—"a spirit which also in emotional terms deserves the name mechanization" (ibid.: 33). Man had become both "machinist and machine" (ibid.: 70).

Modern society, in Rathenau's analysis, was an instrumental society in which everything—including fellow human beings—was reduced to tools. Modern man was a "man of purpose," not a "man of soul" (1918g: 30, 49). To paraphrase Rathenau's contemporary Max Weber, one might say that Rathenau attempted to show that the world was characterized by an ever-expanding purposive rationality. People in modern society set goals and tried to organize their lives in such a way that they could reach these goals with a minimum of effort. However, their goals were flawed. The relationship between means and ends had been reversed, and "mechanical production [had] raised itself to an aim in its own right" (Rathenau 1918f: 44). Technology and technological development were no longer neutral tools in the pursuit of various purposes; higher productivity and increased production had become goals in themselves.

Some of what Rathenau preached during and after the war was at odds with what he had said earlier. From 1915 on he called for increased productivity, but before the war he had distastefully rejected the modern concern with material production. In 1913 he had not argued in favor of corporate restructuring measures and large-scale industry, but had praised "the artisan of the old kind, who manufactures household items for their own sake, with perfection as his goal" (1918g: 65). He had made the industrial mode of production the focus of his critique in 1912 and 1913, but a few years later he saw the same phenomenon as the savior of Germany.

Rathenau's discussion of the German state went through a similar reassessment. Rathenau had not attributed a great deal of importance to the state before the war. In fact, he saw its role as limited to "foreign policy, national defense, legislation, and legal protection" (1918f: 50). Such a "night-watch state" is, of course, light-years away from the strong, paternalistic state that Rathenau depicted after the war. How can such a radical change of positions be accounted for? The answer, I believe, can be found both in his texts and in the contemporary context. Rathenau did not anticipate the war and its consequences; in 1912 he even thought that all nations would be able to dispose of the military in the not-too-distant future. The problem, in a situation of increasing wealth and productivity, was not how to rationalize industry but how to restrict the impact of the mechanical spirit. New and unexpected challenges came with the war: How to save Germany from military defeat? How to save the German people from economic ruin and even starvation? Rathenau turned to an idea that he had developed in several prewar publications: the concept of Euplutism (1918f: 60f.).

The term "Euplutism" literally means the well-being of the many. Rathenau had used it in 1912 to describe a society in which collective needs are more important than individual needs. In this society there were no monopolies, no inherited wealth, and no unjust distribution of wealth—just as in his postwar vision of the New Society. The main difference is that it took the war for Rathenau to find the carrier of the idea of Euplutism. Whereas before the war he had hoped that an ethical resurrection of communal values would be carried by individuals, during and after the war he placed all his confidence in the state. In doing so, he revived what McNeill (1983: 339) has called "principles [that] had deep roots in the Prussian past." According to McNeill (ibid.), "rulers from the Great Elector to Frederick the Great in moments of crisis had commandeered supplies as needed, subordinating private interest ruthlessly to the collective, military effort."

Rathenau became a symbol for both postwar recovery and the young Weimar Republic (Volkov 1990), and, like the republic, he largely failed. He was too much of an outsider to have the power to affect German society in the direction he wanted. His ideas about how to organize industry were too radical and aroused intense opposition from business circles (Schulin 1990: 64). Simultaneously, his foreign policy was not radical enough to appease the extreme right. In 1922 he was shot to death by a member of an extreme right-wing group called Organisation Consul.

Charismatic Liberalism and Disciplined Technology: Weber's and Sombart's Solutions to the First Crisis of Modernity

In October 1910 the German Society for Sociology held a national conference in Frankfurt am Main. Werner Sombart, a professor at the Berlin Business School, gave a speech with the brief title "Technik und Kultur" (Sombart 1911). It is not known whether he was aware of Mayer's book with the same title or of the struggle between Mayer and Dessauer. It is, however, very interesting to see that the theme of technology and culture had made its way into the center of German intellectual life before World War I. Although the war certainly intensified and changed the content of the technology debate, it did not start the "struggle" about technology.

No less a person than Max Weber took part in the debate that followed Sombart's speech. The German discourse on technology and culture was not limited to a few marginal mandarins and engineers. However, it was formulated differently in sociological circles, and the difference is still obvious today.

Sombart's contribution to the emerging discourse was later published in *Archives for Social Science and Social Policy*, a journal that then edited together with Max Weber and Edgar Jaffé. Herf (1984: 133) dismisses this sober-minded account of the relationship between technological development and cultural life as a "fairly conventional sample of cultural despair," but I feel that it deserves a new reading.

Sombart was concerned about the excessive and unreflective technological optimism he found among most of his contemporaries. In his analysis of the negative consequences of modern technology, he chose to focus on how it affected music: "The progress of technology enables us to make the theater and the concert-halls ever larger and brighter illuminated. No wonder that the music is also adjusted to these new creations, and that it ends up as far away from coziness and 'intimate' familiarity as the rooms in which it is performed." (Sombart 1911: 346)

Sombart also regretted that advancements in transportation had led to global uniformity and to the disappearance of local musical traditions. However, he acknowledged that the existence of modern means of transportation enable people who did not live in cultural centers to enjoy good concerts. "Music machines," he felt, had the same Janus face (Sombart 1911: 347). On the one hand, they spread a terrible noise that "shakes the nerves" of bar visitors; on the other hand, they made music accessible to the average citizen (ibid.: 345). In fact, Sombart suggested that such machines might very well refine and cultivate "the masses," and he did not consider it impossible that they might "lead to a complete revolution of our social and cultural relations" (ibid.: 347).

Sombart's speech was not simply one of cultural despair; it was a very early attempt to lay the foundation for something that nowadays is called *Techniksoziologie*. Sombart did not use that term, but his stated purpose was to propose a methodology and some concepts for the analysis of the complex relationship between *Technik* and *Kultur*. He did so by presenting a number of definitions and by giving some examples of how such an analysis could be carried out. Sombart's discussion of technology and music appeared at the end of the speech and was intended to illustrate his approach in a "paradigmatic" manner (ibid.: 342).

Although Sombart's primary topic was the dependence of culture on technology, it is clear that he was very much aware of the reverse relationship. A complete analysis would have to determine "how a particular cultural trait . . . influences the quantitative and qualitative formation of technology . . . and . . . how, in certain areas of culture, interests emerge that have an impact on technical developments" (ibid.: 313).

Sombart mentioned a large number of social factors that affected the content and direction of technology, including the policy of the state, the level of scientific knowledge, the degree of religiosity, and the demands of the workers. The impact of these factors was most obvious in the economic sphere, and Sombart rehearsed the famous difference between the handicraft and the capitalist system when it comes to interest in technical change.

Quite naturally, it would be anachronistic to call Sombart a social or cultural constructivist, or an anti-technological-determinist (see Smith and Marx 1994), even though his thinking contained strands that we can put into both camps. It is interesting to note that he categorically rejected Karl Marx's "technological view of history" (Sombart 1911: 314). Sombart quoted a passage from *Capital* and concluded that Marx regarded the productive forces to be the ultimate roots of all economic, political, and spiritual life. Sombart claimed that Marx wrongly saw technology as an independent variable, and went on to say: "No technology exists in a (socially) empty space; it is also impossible for a technology to exist in an Archimedian point outside of human culture, and to exercise influence from there. The interaction theorist is (in this case) correct when saying that 'Everything causes (and influences) everything.' " (ibid.: 316)

For Sombart, technology was not an autonomous realm of society. Instead, he talked about the relationship between technology and other spheres of society in terms of a "cultural carpet" (ibid.: 311)—a concept that sounds strikingly similar to Hughes's (1987) "seamless web."

There is another, perhaps curious, parallel to be found between Hughes and Sombart: both of them talk about technology in terms of temporary or geographically delimited "styles." Whereas Sombart discusses what he called the "cultural style" of technology, Hughes (1987) uses the concept of "technological style" to designate regional or national differences in the design of technical systems or artifacts. In trying to find the core characteristics of modern technology, Sombart

(1911: 309) argued as follows: "The technology of our times is rational, in opposition to earlier technology that was empirically oriented; . . . it . . . goes beyond the limitations imposed by organic nature. . . ." If we want to understand the character of modern technology, he said, we have to inquire into the demands posed upon it by capitalism. Indeed, Sombart claimed that "it is the very nature of capitalism to strive for rational technology, mechanical technology, technical progress" (ibid.: 314). This did not mean that modern technology is a direct function of the capitalist economic system, but it implied that capitalism tends to foster a rational type of technology and a particular style that contains a large amount of scientific knowledge.[10] No doubt, Sombart's terms of analysis have much in common with those that Weber would later use in his speech "Science as a Vocation" (Weber 1919) and in his *General Economic History* (Weber 1923) (cf. Hård 1994b).

Sombart received much support from Weber in the ensuing discussion, especially for his critique of Marx. Marx's materialist view of society and history had long been, and would remain, one of Sombart's and Weber's main targets (Krause 1960; Mitzman 1971). Weber (1924: 450) not only claimed that Marx's famous technological determinist statement that the handmill creates a feudal society whereas the steam mill creates a capitalist society "is simply false"; he also attacked those contemporary engineers who seemed "seriously to believe that technological evolution is the ultimate driving force in the development of culture" (ibid.: 451f.). Like Sombart, Weber (ibid.: 456) went on to argue that it is, in principle, impossible to find any final cause in human history:

I would like to lodge a protest against . . . the notion that something—be it technology or economy—can be the "last" or "final" or "true" cause of anything. When we analyze a causal chain, we always find that it runs first from technical to economic and political things and then from political to religious and economic things. Nowhere do we find a point of rest.

This Weberian anti-determinism—by no means limited to the realm of society and technology—has been observed by many scholars. For instance, Collins (1986: 25) has said that "in Weber's scheme, technology is essentially a dependent variable." (Cf. Abramowski 1966; Schmidt 1981; Schluchter 1989.)

10. By this time Sombart, like Dessauer and others, had begun to expand the connotation of *Technik* beyond mere machines.

Weber (1924) came back to Sombart's example of the relationship between music and technology, suggesting that modern instruments had, to a large extent, been developed in direct response to the needs expressed by composers. Here Weber implicitly attacked Sombart for having put too much emphasis on the negative impacts of technology on musical forms.[11] Weber (ibid.: 455) claimed that it was incorrect to say that the music of a certain time period was "a product of the technical situation."[12] In the case of chemistry it might be possible to talk about a direct dependence on technical and economic developments, but in the case of music it would be more appropriate to say that the composer made use of whatever technological possibilities were at hand and acted within their limits. In short, both Weber and Sombart went out of their way to defend an irreductionist view of history in general and an anti-determinist view of the relationship between technology and culture in particular. They agreed that technological developments affected social life, but they never claimed that this was the whole story. In short, they picked up themes that are still at the core of the social studies of technology, contributing to the foundation of a discursive framework for this academic field of investigation.

Wagner (1994: 68f.) describes National Socialism as an ideology that, first and foremost, attempted to organize modern life. Referring to Bauman 1989, he concludes that, although the Nazi movement went to extremes in this endeavor, it must not be considered different in principle from "Stalinism, People's Front (and later Vichy), People's Home and New Deal." These movements or projects felt uneasy with the liberal utopia of classical modernity and with its emphasis on individual freedom and *laissez faire*. Instead, they wanted to control the economy for the benefit of the many, steer science in a socially and politically acceptable direction, and put the collective in front of the individual. For present purposes, the interesting thing is that they all made extensive use of up-to-date technology in these endeavors. In other words, they applied modern means to the control of modern life.

It is probably no coincidence that the term "technocracy" reached wide popularity in the heyday of the above-mentioned movements and

11. It seems to me that Weber is somewhat unfair here. Sombart had made it clear very early in his speech that he would focus on technology's impacts in order not to make his speech too long. He had also said that "music life is connected to technology with thousands of threads" (Sombart 1911: 342).

12. As far as I can tell, Sombart had never claimed this.

projects (chapters 4 and 5 below; Layton 1971, chapter 10) and that concerted attempts to create a kind of "social engineering" were made at this time. Many engineers and social scientists felt in the 1930s that the time was ripe for their expertise to influence politics. In Germany the Nazis' seizure of power was regarded by several engineers as a heaven-sent opportunity to affect the development of society at large. Renneberg and Walker (1993: 8) observe that "technocratic enthusiasts flooded into the [Nazi] Party or ancillary organizations after 1933." Some of these enthusiasts went a little too far in their attempts to gain influence, however, and in 1935 the Nazis ordered the closing of the journal *Technokratie*—"ironically just when opportunities for technocrats within the National Socialist state began to improve" (ibid.: 5). Renneberg and Walker (ibid.: 9) agree with Herf's (1984) assertion that the Nazis were not generally opposed to modern technology, and contend that "the common assumption that the National Socialist movement deliberately set out to purge science or engineering in particular is questionable."

Even though this "common assumption" is usually based on a few cases (including the Nazis' attempts to curb post-Newtonian physics and their inability to appreciate the potential of the atom bomb), it does seem understandable. Indeed, the Nazis' view of technology was not very clear, even to their contemporaries. One year after the National Socialist party came to power, Sombart published a book that vindicates this observation. His *Deutscher Sozialismus* (1934), which must be considered an overt flirtation with the new government, included an elaborate plan for how modern technology should be controlled and how science should be disciplined (cf. Herf 1984: 148ff.). Sombart's ideas did not fall on fertile ground. A Nazi reviewer of the book wrote distastefully: "The National Socialist position vis-à-vis technology has nothing whatsoever to do with Sombart's. . . . Modern technology is for us a child of the Northern spirit, and expresses the power of our mankind."[13] Obviously, Sombart had misinterpreted the Nazi ideology on this score.

Sombart viewed modern technology more critically than did the Nazis and Rathenau. With his ideas about the "taming of technology" he did not merely want to organize modern society along rational lines; he also wanted to reject certain outcomes of modernity as such (Sombart 1934: 264). Sombart had already attacked the notion of autonomous technol-

13. Review by a certain Nonnenbruch in *Völkischer Beobachter* (no. 278, October 3, 1934); quoted from Krause (1960: 189f.).

ogy immediately after World War I (Sieferle 1984: 215), and in his book *Die Zukunft des Kapitalismus* (1932) he had, in Krause's (1960: 175) words, argued in favor of "an orderly, tamed, domesticated, meaningful, and organized economy." In 1934 he finally felt that the time had come to put forth a full-fledged program for how the development and use of inappropriate technologies should be checked: "Each invention has to be approved by a supreme cultural council [*oberster Kulturrat*], where technologists have an advisory function. The cultural council decides if the invention should be discarded, handed over to a museum, or implemented." (Sombart 1934: 266) It is no wonder that Sombart irritated engineers and others who had technocratic inclinations.

In their analysis of the interwar "regulation" discourse, Dierkes et al. (1988: 13f.) refer explicitly to Sombart's "German Socialism." Sombart's ideas about how to tame technology must be considered an extreme manifestation of this discourse, which we nowadays find in various discussions about how to "regulate," "assess," and "evaluate" technology. He not only suggested that the government appoint a body to decide what new inventions should be developed; he also called on the police to prohibit the use of those existing technologies that disturbed the common citizen or threatened workers' health, and referred with approval to the earlier decision of the Swiss canton of Graubünden to ban the use of automobiles and motorcycles. As a final point in his plan to control technology, he called for the foundation of a government research institute, on the ground that the state and not the market should decide in what directions scientific and technological research should move.

The main reasons for Sombart's suggestions are to be found in his dislike of unfettered capitalism. Sombart had never reconciled himself fully with liberalism, and in his later writings he grew increasingly critical of the liberal market economy (Harris 1958, chapter 6). No doubt the German experience with ultra-high inflation in the early 1920s and with economic hardship around 1930 triggered these reactions. In the third volume of his monumental work *Modern Capitalism* (1927) he prophesied the downfall of capitalism, and in *Die drei Nationalökonomien* (1930) he sought to substitute a new economic theory for the narrow neoclassical theory. In *The Future of Capitalism* Sombart argued in favor of a kind of organized capitalism with cartels and monopolies instead of short-sighted, cutthroat competition. Responsible planning, he argued, ought to replace egotistic profit maximization. In 1934 he was ready to develop the consequences of this thinking for the area of technology. Technological development, he wrote, should not be guided by profit motives and private business interests (Sombart 1934: 257). Instead of passively observing how "the formation of our material culture is at the mercy of a group of inventors and smart business people" (ibid.: 264), the state should actively intervene in the development process. Only then would it be possible to "put order into the chaos" (ibid.: 266). If technology were left to the market forces, it would most certainly continue to develop in a direction that did not serve the many.

Sombart's formal definition of technology did not change between the early 1910s and the 1930s. *Technik* was "all systems (complexes, totalities) of means, that are suited ... to fulfill a certain purpose" (Sombart 1934: 245).[14] "Technology" retained its instrumental definition, but in 1934 Sombart was more explicit about the consequences of this definition. First, it meant that "technology is always culturally neutral and morally indifferent: it may serve either the good or the bad" (ibid.: 262). Second, it meant that technology cannot have any inherent value. It was, for instance, impossible to construct a scale by which different technologies could be measured (ibid.: 252).

Against this conceptual background it may, of course, be asked what criteria Sombart's cultural council could have used when approving or rejecting various inventions. Sombart never seemed to acknowledge this problem. Instead, he chose to attack those of his contemporaries who had turned the means-end dichotomy on its head and made technology an end in itself. The phrase "l'art pour l'art" was, he wrote, symptomatic for the whole "technological age" (ibid.: 255, 252). Technology has reached such high esteem that everything that was possible to make also appeared to be worth making, "without asking about the purposes that it should fulfill" (ibid.: 254). This unreflective and dangerous view led to undue respect for "the engineer and his ways of thinking: the word 'technocracy' could only appear in our age" (ibid.: 253).

The liberal economy was thus not the only target of Sombart's critique. His call for an institutional form of technological assessment can be read as a reaction to what he saw as an increasingly aggressive engineering corps. Although Sombart's solution was clearly elitist, it denied the engineers the final say. Like the "leader council" (*Führerrat*) that was to get the last word in all important political matters (Sombart 1934: 214), the cultural council was to be a body of first-rate experts and politicians that

14. The only difference between the 1911 and the 1934 texts is that the word "totalities" was added in the latter.

would fit well into the contemporary regulatory discourse. Like so many others in the first 40 years of the twentieth century, Sombart tried to *organize* modernity.

Sombart's diagnosis for reversing the means-end relationship echoed something that Weber had begun to discuss 30 years earlier. Weber's famous phrase at the end of *The Protestant Ethic and the Spirit of Capitalism* (1904–05) about "the iron-cage" of the modern, materialist world was meant to depict a situation where the originally purposive-rational means of technology and bureaucracy had reached autonomy and begun to threaten the freedom of the individual (Habermas 1981, I: 337). Instead of being means, the instruments tended to become ends in themselves.

Like Sombart, Weber (1922a: 160) gave technology an instrumental definition: "The term 'technology' applied to an action refers to the totality of means employed as opposed to the meaning or end to which the action is, in the last analysis, oriented." Bureaucracy was discussed along similar lines: "The decisive reason for the advance of bureaucratic organization has always been its purely technical superiority over any other form of organization. The fully developed bureaucratic mechanism compares with other organizations exactly as does the machine with the non-mechanical modes of production." (Weber 1922b: 214)

The key terms in Weber's analyses of technology and bureaucracy were "predictability" and "control" (Collins 1986). The scientific basis of modern technology and the bureaucratic structure of modern organizations made it particularly easy for the industrialist to predict the outcome of the production process and for the corporate director or the minister to control his employees or officials. However, the contemporary problem was that these structures showed a tendency to become increasingly self-contained as they grow in size. The more powerful the tools, the more difficult to govern.

Weber experienced both the emergence of the political party apparatus and the growing power of cartels and trusts, and he interpreted these processes in terms of the bureaucratization of politics and business. Despite Weber's attempts to produce "objective" and disinterested science, it is clear from his writings that he bemoaned these developments (Abramowski 1966: 161ff.; Marcuse 1968: 224f.). Weber regarded the growth in the Occident of this kind of rationality as "tragic" (Åmark 1990), and he "showed himself to be critical of the extension of formal rationality as an end in itself" (Beetham 1985: 274). The solution he offered to the means-ends problem was partly similar and partly different from Sombart's.

Like Sombart, Weber wanted clearly stated values and ideologies to influence not only politics but also business and technology. Neither of these areas of modern life should be excluded from what Weber (1922a: 115, 185) called "value rationality" and "substantial rationality." That is, it had to be possible for values such as "national grandeur" to bear on politics and, at least in principle, for "equality" to influence the domain of business. As Beetham has indicated, though, Weber's value-neutral methodology forced him—at least in his scientific writings—to postpone more elaborate discussions about how the capitalist system could be made more substantially rational. Furthermore, Weber's ideological leanings led him to believe that the economic efficiency of capitalism would, in the end, make it easier for a capitalist society than for a socialist one to improve the conditions of the poorer classes: "Despite all the bureaucratization of modern society, which was itself a consequence of capitalism, Weber still believed that the accumulation of profit in 'rational' undertakings could be given an ethical significance." (Beetham 1985: 273) Unlike Sombart, who disliked pure profit motives, Weber remained too much a bourgeois liberal to put his faith in any economic system other than capitalism (Mommsen 1984/1974).

In another respect, however, Sombart and Weber were not far apart. Both called for a powerful state and a strong leader, although for different reasons. Whereas Sombart saw the state as a neutral power that could mediate between the classes, bring social ideals into capitalism, and tame technology (Krause 1960), Weber wanted a united Germany and a potent state primarily for reasons of international prestige and strength (Mommsen 1984: 394). Whereas Sombart (1934: 224) did not hesitate to call for "a total social order within the state," Weber was afraid that the expansion of state influence into too many areas of social life would further reinforce the iron cage. Weber's aversion to social democrats and communists was in part based on what he found to be their predilection for bureaucratic structures. In order to get out of the cage, it was necessary to put faith in a strong personality rather than in a cultural council and to turn to someone with "charismatic authority" rather than to a strong bureaucracy (Weber 1922c). A charismatic person would be able to break with old traditions and to lead people and organizations onto new revolutionary paths by his sheer "specific and exceptional sanctity, heroism or exemplary character" (Weber 1922a: 328).

Weber's faith in the persuasive powers of certain individuals was not opposed to his belief in the state. The existence of an effective state was necessary in order for the idea of the charismatic politician to materialize and gain legitimacy. In much the same way that a machine can continue to be used even if a factory receives a new director or owner, a well-organized bureaucracy should be able to serve a new lord.

Conclusion: The Mobilization of Tradition in Periods of Crisis

We have in this chapter encountered several intellectual reactions to the problems associated with the first crisis of modernity. Two interconnected responses stand out. One made recourse to the notion of German culture and one to the German state, and both share a strong nationalist bias. I would like to argue that German intellectuals by and large tried to come to grips with the threats of modern technology by taking refuge in nationalist ideas of a superior German culture and a powerful German state. In their formulation of the German *Sonderweg*, both *die Kultur* and *der Staat* should save the German people from the most devastating effects of mechanization and Americanization (Jakobsen et al., chapter 5 in this volume; Kocka 1988). There were, of course, cosmopolitans, liberals, and futurists in Germany, but a strikingly large segment of German intellectuals chose to meet modernity by returning to traditional but redefined ideas of German particularity. In so doing, they attempted to assimilate modern technology into German culture, thus domesticating it. They tried to make modern life acceptable by carving out an appropriate place for it within existing structures, and by arguing that the German state had to play a central role in this process.

Dann (1993) has shown how the idea of a united German realm or nation has been a powerful constituent in German thinking from the late eighteenth century onward—with traces back to the Holy Roman Empire of the German Nation. The idea can be found among groups of various political coloration and has taken on partly different meanings in different circumstances. It has appeared under several disguises, each with a slightly particular connotation: *Reichsnation, Nationalstaat, Kulturnation,* and, of course, *Dritte Reich.* I suspect that the flexibility of this notion explains why it has been possible for intellectuals and politicians from various ideological camps to redefine it and to make use of it in new situations.

Weber and Sombart picked up this German tradition in their positive evaluation of the state. Both Weber's parliamentary-democratic state under charismatic rule and Sombart's elitist German Socialist state built on ideals that had a long history in their country. Similarly, Rathenau's whole program of postwar restructuring was nationalist at root. His ''New

State" was in a sense not very new at all, but rather a reformulation of ideals that had been held in high esteem in Prussia since well before the days of Bismarck. Finally, mandarins such as Mayer and engineers such as Weihe went out of their way to reinvent the old notion of German culture so that it could accommodate and encompass modern technology.

The story does not end here, though. As was indicated at the beginning of this chapter, the modes of addressing the question of technology that developed in the first decades of this century became paradigmatic for at least three different discourses. First, the originally German version of *Technikphilosophie*—which after World War II would develop further in the United States—received momentum in this period. For example, in *Technological Enlightenment* (1991) the philosopher Günter Ropohl deals with several of the same themes and makes use of several of the same concepts as Dessauer and Weihe. Second, the tongue-tying notion of *Technikfolgenabschätzung* received its cognitive basis in this period. When the German authorities evaluated the future of the former East Germany's nuclear power plants a few years ago, they acted much like Sombart's cultural council, deciding whether or not an alien technology should be allowed to be integrated into the "New State" of reunified Germany (Dierkes et al. 1988). Third, the fashionable academic field of *Techniksoziologie* can in certain respects trace its roots back to the early decades of this century. When the German sociologist Werner Rammert (1993) argues against technological determinism and in favor of an understanding of technology as a social and cultural product, he echoes not least Weber's way of posing his research problem (Collins 1986; Hård 1994).

The argument should not be misconstrued. The fact that the discursive frameworks were set during the period under discussion does not mean that the debates have remained the same ever since then. As Peter Wagner points out in chapter 9 below, new contingencies and contexts make new understandings possible. Traditions are not only straitjackets; they also give actors leeway and freedom. However, it is presumably no coincidence that the three discourses mentioned here have such strong positions in Germany, where they have been formed and reformed continually for almost 100 years.

4

American Anxieties: Technology and the Reshaping of Republican Values
Andrew Jamison

Setting the Stage

Any attempt to summarize in a relatively brief space the intellectual discussions about technology in the United States is bound to be somewhat arbitrary in its coverage, so let me explain the main motivations behind my particular selection of material. My primary aim has been to survey the recent historical literature in order to provide a basis for comparison with the other chapters in this volume. Another aim has been to identify intellectual positions in relation to modern technology that can help inform contemporary discussions. Indeed, I would argue that contemporary discussions cannot adequately be understood unless we return to the early years of the twentieth century, when American intellectuals, in the wake of World War I, started to become aware of modern technology as a significant social problem and to develop frameworks of interpretation for "appropriating the machine."

Although they were shaped by their historical context, the positions that were staked out in the interwar years by Thorstein Veblen and Lewis Mumford, in particular, continue to influence the ways in which Americans think about technology. Technocratic disciples of Veblen, be they institutional economists, corporate managers, or technology planners or policy makers, continue to do battle with Mumford-inspired critics—the disparate cluster of historians, philosophers, and environmentalists who seek to humanize the machine—even though the original sources for their viewpoints and disagreements may not be readily apparent. A closer look at the debates of the early twentieth century can thus provide an important historical perspective on the debates that are taking place as the century comes to an end.

In his survey of American technological history, Hughes (1989: 249) has noted that "the United States had never enjoyed greater respect, or been more envied, than after World War I. Many foreign liberals and radicals perceived its examples as opening for their nations a path to the future." It was particularly the new production technologies and organizational techniques—the system of production—that inspired admiration; the social system, and the cultural ideals that guided the new technologies were often seen more critically. Indeed, there was concern that the new mechanical civilization in the United States would lead to a general decline in cultural values. Max Weber, Martin Heidegger, José Ortega y Gasset, and other European "mandarins" emphasized what was lost, or left behind, in the promulgation of modern instrumental values, and these attitudes were to find a certain resonance among younger intellectuals in the United States, such as Lewis Mumford and Joseph Wood Krutch. Interestingly enough, it would be largely under the influence of such ideas as they came to be expressed by transplanted German intellectuals, emigrating to the United States to escape Nazi oppression, that a critique of "mass culture" would develop in the United States in the 1940s. Herbert Marcuse, Hannah Arendt, Theodor Adorno, and Erich Fromm, among others, were important carriers of the German interwar technology debate to the postwar United States. These disciples of Heidegger and Weber would criticize the hold of technological rationality over American life and identify new forms of authoritarian control based on instrumentalism, which Marcuse came to call "one-dimensional thought" (Jamison and Eyerman 1994). In the mixing of this imported "critical theory" with the more indigenous brands of reflection that I discuss in this chapter, the intellectual seeds were sown for a new wave of technology debates that have emerged in recent decades.

In the 1920s and the 1930s, American intellectuals sought to "reinvent" their country as a global technological leader and as the embodiment of a new kind of civilization. Although there were criticisms of the technological regime that took shape in the United States, the interwar years generally were a time in which the new technological potentialities were seen not so much as threats as opportunities. "The American beams with a certain self-confidence and sense of mastery," the philosopher George Santayana—himself an immigrant from Spain, who eventually resettled in Italy—proclaimed in 1918. "He feels that God and nature are working with him." (quoted in Tichi 1987: 120) Fordism and Taylorism, derived from Henry Ford's methods of mass production and Frederick

Winslow Taylor's techniques of scientific management, led many in the United States to celebrate that country's emergence as an international technological and economic leader. After the war, American styles of production and consumption spread around the world as the quintessential models of modernity, even in the new Soviet Union. As Stalin put it in 1924: "American efficiency is that indomitable force which neither knows nor recognizes obstacles; which continues on a task once started until it is finished, even if it is a minor task; and without which serious work is inconceivable" (quoted in Hughes 1989: 251).

The Discursive Framework: Technology and Republican Values

American culture has long been identified with technological progress. Among the earliest colonists were Puritans, who brought an interest in Baconian "useful knowledge" and technical improvements with them from Britain in the seventeenth century. John Kasson, in his book *Civilizing the Machine* (1977), points to the role that things technical—the so-called mechanical arts—played during the revolutionary period, as practical ingenuity and a search for labor-saving devices came to form a central part of what he and other historians have termed republican ideology. Above all else, the American was an industrious and skilled worker, transforming (either as farmer or artisan or homemaker) a bounteous nature into useful products. Indeed, the memory of a "republican technology" continued to provide important cultural resources for those who would appropriate the machine into twentieth-century American civilization.

Republicanism served as an idiomatic filter, translating commonsensical everyday beliefs into a more coherent framework of ideas in which individual freedom, equality of opportunity, American exceptionalism, and technological progress were combined into a collective social vision (Ross 1991). In the garden that was America, there was plenty of room and understanding for the machine; according to the influential account of Leo Marx (1964), American culture assimilated technology through the discursive framework of pastoralism. The machine, conceptualized as a helpful mediator between nature and society, was generally accepted as one of the main contributors to progress, which in the American context came to mean the conquest of the continent, the exploration of the frontier, and the taming of the wide open spaces and native peoples of the American West.

With the closing of the frontier, science and technology—and the positivist philosophy they had stimulated—came to be seen as the new arenas of expansion and modernization (Wilson 1990). The main task was to diffuse their fruits among the population and, by so doing, integrate the machine into the American republic. There was a general acceptance of the machine, and only a small number of "genteel" intellectuals took on the task of upholding a sense of Old World morality in an all-too-vulgar society (Perry 1989). But even the genteel believed in progress; their main complaint was not with technology but with the anti-intellectual climate that a hegemonic culture of practical and commercial interests so readily fostered (Lears 1981; Leach 1993). Where opinions differed was in regard to who should determine the course of the diffusion process: should it be the farmers, with their ambiguous populism and their dedication to hard work; the engineers, with their mechanical logic and technical rationalism; the "captains of industry," the capitalist barons who had brought the machinery into productive use; or the captains' opponents in the increasingly militant working class?

The United States approached the twentieth century as a young nation moving rapidly toward the center of the international stage. With the growth of powerful industrial empires in steel, oil, chemicals, telephones, electricity, transportation, weaponry, and machinery, the United States had come to define the very meaning of technological progress, having taken the poor and dispossessed immigrants from Europe and created a dynamic, productive nation. Of course, the impact of industrialization had been stratified; many had benefited, but many had suffered and had sought, in the course of the nineteenth century, to challenge some of the more noticeable problems that had arisen. Particularly influential in the political arena were populists of various stripes, who opposed the new industrial elites of managers and bankers with what might be termed democratic alternatives. Henry George's book *Progress and Poverty* (1879), with its support for small industry, had given some intellectual shape to the populist movements, which provided an option for many of those who saw a threat to democracy and republican values in the emerging corporate order (Lasch 1991: 63ff.).

There were regional and ethnic as well as class differences in the ways intellectuals responded to industrial development and to modern technology; however, it is usually argued that most Americans were positively disposed to the machines, which were given a central place in the national identity (Boorstin 1987). The symbolic importance attached to Benjamin Franklin, the Philadelphia artisan printer and scientist who discovered the electrical current in lightning and carried out sensitive diplomatic missions during and after the Revolution, lived on after the new republic had been formed.

In the middle of the nineteenth century, Ralph Waldo Emerson equated technological dynamism with American character. Even though Emerson was critical of technological development that was not controlled by human beings, he welcomed the opportunities that machinery provided. "All the elements whose aid man calls in will sometimes become his masters, especially those of most subtle force," Emerson wrote in 1860. "Shall he then renounce steam, fire and electricity, or shall he learn to deal with them? The rule for this whole class of agencies is,—all *plus* is good; only put it in the right place." (Emerson, *The Conduct of Life*, quoted in Kasson 1977: 134) Emerson's friend Henry David Thoreau left urban life to build himself a house near Walden Pond in order to return to a more natural existence, thus initiating what would be a significant romantic current in American arts and letters, but even he did not really reject technology (Marx 1988). Throughout the nineteenth century, it seems safe to say, Americans, far more generally than their counterparts in Europe, were infatuated with machinery, which, in a country of wilderness and limited manpower, seemed to provide the main vehicle for national greatness. As Molella (1990: 23) puts it: "While Europeans appreciated technology, the love of it was considered essential to the American character." Not until World War I would more discordant tones enter into the peculiarly American love affair with technology.

The war divided intellectuals among themselves and encouraged a sense of disillusion among many of those who had earlier believed not merely in the promise of a peaceful industrial future but also in the general beneficence of the American government. In the words of Lasch (1991: 107): "For those who lived through the cataclysm of the First World War, disillusionment was a collective experience—not just a function of the passage from youth to adulthood but of historical events that made the prewar world appear innocent and remote." For some, pre-industrial values were counterposed to the dominant instrumentalism and to the apparent lack of any social control over technology. Longing for earlier times—classical antiquity, the Middle Ages, even the more remote past of the so-called primitive peoples—became one kind of response. Many young American intellectuals went to Europe, partly in search of older values and partly just to get away. For other intellectuals, the response was to propagate one or another "modernist" ideal of culture based on the methods and products of scientific-technical reason.

This modernism was seen as portending a new historical epoch, and the main task for the intellectual was to give this scientific-technical rationality a larger influence in the body politic and in the world of "high" culture.

For many, the machine itself seemed to be in charge of social development without having any particular goal in mind. The task was thus to define goals for the mechanical civilization derived from the "traditions" of the American past, and, in particular, from the characteristic virtues of republicanism: equality, frugality, public service. For others, archaic social institutions and ways of life were seen as constraining the machine from unfolding its logic in a comprehensive way—and a new kind of republicanism, one more subservient to the demands of technological efficiency, was therefore required. Thus, the lines of the debate came to be drawn between those who would try to "assimilate" the machine into existing social and cultural patterns and those who would seek to "adapt" society and the human personality to the imperatives of the machine age. The debate about technology in the 1920s thus often took the form of conflict, or tension, between the articulators of one or another set of cultural values on the one hand and the defenders of an instrumental civilization of technically-oriented material progress and economic prosperity on the other. In the pages of new intellectual journals, such as the symptomatically entitled *New Republic* (founded in 1914), intellectuals took on the task of providing technical development with a new kind of legitimation: a new form of "republican" virtue that would be appropriate for the twentieth century.

Both in reality and symbolically, the war marked the end of the so-called progressive era, with its active programs of government reform and its even more active flowering of radical politics and social movements. The debates about American involvement in the war, and then the divergent assessments of the war's implications, served to break up what had been a progressive consensus, grounded in the pragmatism of William James and John Dewey and epitomized by the social work of Jane Addams and her Hull House in Chicago. In the 20 years that preceded World War I, a broad movement of politicians, intellectuals, and academics had sought to improve the social order and bring some of the more problematic aspects of industrial society under control. During Theodore Roosevelt's presidential administration, "muck-raking" journalists, city and state government officials, and public service organizations manifested a sweeping spirit of reform, among the goals of which were to break up the trusts, to help the poor and downtrodden, and to bring planning and expertise into American public life.

This was the age of the pragmatic expert, who took on important new social functions in technological development and in politics. Except for a few disenchanted old aristocrats, such as Henry Adams, the progressive era marked a new period in the continuing celebration of modern technology. The new century was to be the American century, and above all else that meant a century of technical improvements. Thomas Edison, Henry Ford, Alexander Graham Bell, and the Wright brothers, the key inventors of modern technology, were all Americans. It would be through ingenuity and inventiveness that the United States would conquer the world (see Hughes 1989). And for a time, the pragmatic philosophy of James and Dewey combined with the future-oriented and reform-minded politics of Theodore Roosevelt and Robert Lafollette (the Progressive Party governor of Wisconsin). The liberal synthesis of the progressive era would be short-lived, however. It could not permanently unite the disparate and often conflicting interests it sought to represent, nor could its proponents agree on the morality of America's participation in the war. By the late 1910s, the progressive synthesis had split apart into a number of different positions.

Progressivism in the United States represented an attempt on the part of the established order to bring modern life in general and capitalism in particular under control. (See chapter 9 of this volume.) Progressives believed in technological and scientific progress, and they generally sought to use "the machine" and its logic to impose some kind of regularity and order on the chaotic society in which they lived. The progressive prophets Thorstein Veblen, John Dewey, William James, Jane Addams, and Charles Beard were all imbued with republican virtues, democratic leanings, and instrumental values. Their faith was generally less ideological than similar political and intellectual currents in Europe—less a program than a vague and somewhat elusive project of reform, which later historians have had some trouble pinning down. (See, e.g., Susman 1984.)

Historical portrayals of the progressive era have tended to reflect the concerns of the historians' own times. The period was given the label "progressive" post facto, after whatever had been accomplished had been transformed into something else (Rodgers 1982). Hofstadter (1955: 5) notes that "it was not so much the movement of any social class, or coalition of classes, against a particular class or group as it was a rather widespread and remarkably good-natured effort of the greater part of society to achieve some not very clearly specified self-reformation."

In the 1960s, Robert Wiebe focused on the middle-class urban basis of progressivism. As Wiebe (1967: 166) put it, in what was to become a standard textbook treatment of the period: "The heart of progressivism was the ambition of the new middle class to fulfill its destiny through bureaucratic means." For Noble (1977: 64), progressivism was an attempt to design a new kind of "corporate capitalism": "At the turn of the century, these reformers embarked upon a far-reaching enterprise, to bring American society into line with technological advance and corporate growth."

Ross (1991: 144–145) has stressed the "pluralistic politics" of the progressive movement, which, "though weighted toward business interests," led to "a series of legislative reforms, including the extension of economic regulation, modernization of the banking system, passage of the income tax, and the first hesitant steps to bring social welfare issues to the attention of the federal government." Ross (ibid.: 143) notes that the leading intellectuals of the period "struggled with socialism and tried now to escape its ideological polarization," and that "the paradigms they formulated—neoclassical economics, liberal economic interpretations of history, a sociology and ideology of social control, and pragmatism—laid the groundwork for twentieth-century social science."

After the war, the prewar ideas lost much of their power and influence as a new generation sought to create its own social criticism and its own intellectual practices. As Randolph Bourne, one of the more vocal public intellectuals of the progressive era, expressed it: "To those of us who have taken [John] Dewey's philosophy almost as our American religion, it never occurred that values could be subordinated to technique." (quoted in Blake 1990: 161) This American version of organized modernity (Wagner 1994) was challenged by the experience of the war. Its hopeful view of the future, with experts in control and with the state providing a paternal helping hand for the private pursuit of profit, seemed for many younger intellectuals after the war to have been little more than a naive dream that needed to be replaced by more hardnosed approaches to the problems of industrial civilization. The 1920s were thus a time for reassessing and eventually redefining the progressive vision.

There was a highly visible shift in the 1920s from a collective concern for social welfare and public service to the more private and individual concerns of the marketplace. The active governments of Theodore

Roosevelt and Woodrow Wilson were replaced by the corporate expansion and consumerism of the "roaring twenties." The corporations served as the main developers of technology, as well as the main sponsors of scientific research and management education. Through the largesse and the policies of the corporate foundations, a particular kind of instrumental expertise, at once quantitative and scientistic, was given a major role to play in American intellectual life. The social sciences, after being associated before the war with the progressive ideas of John Dewey and Jane Addams, were largely transformed into instruments of corporate development (Ross 1991).

The 10 years that followed the war were years of rapid technological change. The private automobile—especially Ford's Model T and, after 1927, the Model A—led to a new sense of personal mobility and contributed to a transformation of the natural landscape. At the same time, the widespread diffusion of electric appliances transformed the home and helped draw firmer lines of demarcation between the public and the private sphere. And there were the movies and radio and other "mass media" to transform the experience of life itself. Sound, sight, and emotion could be mechanically recorded, reproduced, and widely disseminated with far-reaching consequences on the common culture. This "culture of consumption," as it has been called, was largely constructed by corporate interests, which established modes of producing, distributing, advertising, and, in the case of the communications technologies, utilizing the new products without any conscious plan or vision of moral enlightenment (Fox and Lears 1983). A technological revolution took place, one might say, without any corresponding revolution in human consciousness. The new technical potentialities were contained within an economic system that was interested only in maximizing profits.

The Cast of Characters

It seems particularly appropriate to approach the technological debate of the 1920s from a cyclical theory of social change in which relatively brief and intensive intervals of "social movement" are seen as being periodically followed by institutional consolidation and incorporation (Eyerman and Jamison 1991). Ideas or intellectual positions which are combined in social movements are decomposed in periods of incorporation. In this way, social movements serve to recombine ideas that, in "normal" times, tend to be formulated by intellectuals linked to disparate, even opposed, social groups and/or political projects. From such a perspective, the 1920s can be seen as an interval between movements, that is, between the labor activism and "progressive" movement activity of the prewar years and the mass mobilizations that would come in the 1930s with the Great Depression. In the 1920s, ideas about technology that can be said to have been recombined for a time in the 1930s were formulated in a differentiated manner by different intellectuals and taken in different ideological and political directions. The 1920s were a time when the ideas of one movement—progressivism—were transformed into the intellectual seeds of another. A key element in this process of intellectual transformation was the reinterpretation of republican ideas about technology.

In what follows, I will briefly examine four positions, which, somewhat schematically, I will label the *technocratic*, the *traditionalist*, the *humanist*, and the *pragmatic* position. The basic distinction is between those who would seek to adapt society to the imperatives of the machine—the technocrats and the pragmatists—and those who sought to assimilate technological development into one or another cultural framework. My slicing of the debate obviously draws on other historical accounts of the period, but the positions are largely of my own construction, since they are meant to lend themselves to comparison with similar debates elsewhere. The positions are meant to capture the range of debate in general terms, as well as to highlight positions that have continued to inform the intellectual appropriation of technology in the United States. It can be suggested that the basic distinction is more European than American, since the critics were and are more marginalized in the American political culture than in the German, the English, or the French. The lack of a pre-industrial culture and its institutional and ideological legacies has kept the anti-industrial critique from becoming as influential in the United States as it has been in Europe. But perhaps it also explains why Lewis Mumford's more constructive kind of humanist critique has been able to gain a wider foothold among intellectuals in the United States than elsewhere.

For reasons of brevity and clarity, I have let one central figure stand for each position, and I have discussed at greater length two figures— Thorstein Veblen and Lewis Mumford—whose opposing positions continue, more than 50 years later, to influence the American debate about technology. Both Veblen and Mumford also gave rise to more formalized, disciplinary discourses about technology: institutional economics and history of technology, respectively. As sources of inspiration, the writings of Veblen and Mumford helped to carve out intellectual "spaces" that continue to be filled and reconstituted. There were, of course, other contributors to the frameworks of interpretation that Veblen and Mumford were perhaps the first to articulate. In the development of institutional, or innovation, economics, the writings of Joseph Schumpeter, who moved to the United States from Austria in the 1930s, would come to be particularly important; and in the history of technology, other writers—perhaps especially the medievalist Lynn White—would specify the methods of analysis that would be developed in later years. While I focus on particular individuals to represent the different positions, it is important to realize that the individuals were involved in broader contexts of opinion-making and intellectual practice.

The grouping of individuals into particular positions is primarily based on the different definitions or characterizations of "culture" that are opposed or contrasted to technology. Thus, I link the avowed technocrats Howard Scott and Stuart Chase and the economic historians A. P. Usher and Joseph Schumpeter with the eccentric theorist Thorstein Veblen, because they draw on a similar—largely negative—notion of culture as outdated and pre-modern. Their critique is of the non-rational, anti-intellectual American cultural values that act, to their way of thinking, as constraints on the proper and effective modernization of the country, and which limit the control of its destiny by experts and engineers. In order to make use of the new productive forces that had been unleashed in the nineteenth century, it was necessary to understand the internal dynamics of technological change and to adapt the institutional and broader cultural contexts to this revolutionary techno-logic. A modern society would transcend the traditions and superstitions of the past with the instrumental rationality that the machine required.

The traditionalist critics T. S. Eliot and Van Wyck Brooks shared much of the modernism but little of the technological enthusiasm of the technocrats. In the United States, they stood for a new kind of disenchanted intellectual role, challenging the vulgarity and materialism of American life with the cultivation derived from classical cultural virtues (Eliot) or with a reconstructed and revitalized version of American history. While Eliot decried the wasteland of contemporary civilization and spent his life defending and articulating a form of cultural elitism, Brooks sought to mobilize the cultural resources of the American past against the commercial culture of the present (Blake 1990). The American traditionalist critique of the modern was, of course, immortalized in the *Education of Henry Adams*, which was extremely influential in artistic and cultural circles during the 1920s. While many artists and authors sang the praises of the machine age, and poets like William Carlos Williams transformed poetic meter itself into a kind of mechanical rhythmic form—cultural critics of industrial civilization mobilized various cultural traditions to challenge the present (Tichi 1987). Joseph Wood Krutch's 1929 book *The Modern Temper* came to epitomize this position.

In the 1920s, a different kind of critique of modern civilization emerged from ecologically or biologically oriented social thinkers, including Lewis Mumford, Howard Odum, and other "regionalists" (Thomas 1990). Mumford's lesser known activities of the 1920s can be seen as a new form of populism; he defined the community as the antipode to civilization. However, for Mumford and Odum the community or region was not merely a nostalgic source of historical traditions and memories; it was also, and even more importantly, a socio-geographic environment, a conditioning place. Out of the ecological or communitarian critique would come new criteria for technological development and new ideas about diffusing knowledge. This critique's wrath would be directed both against the undesirable social and human consequences of technology and against the overextension of instrumental rationality into modern American life. What Odum termed "super-civilization" stood, as he put it, "in many bold contrasts to culture," putting "organization over people, mass over individual, power over freedom, machines over men, quantity over quality, artificial over natural, technological over human, production over reproduction" (quoted in Pells 1973: 102). For the human ecologists, culture included social traditions as well as natural conditions; both needed to be mobilized to encourage what came to be called a regionalist approach to development.

My fourth category is where I place the critique of Western civilization emanating from the pragmatism of John Dewey and his disciples. In the United States, the pragmatists' critique was of the values of modern civilization, rather than of the technological achievements. In this category one can also place the young Reinhold Niebuhr and other socially involved theologians, who were not critical of technology itself, as much as the powerful capitalists who controlled its development. The call, from Dewey as well as from Niebuhr in the 1920s—and, from a very different direction, from the young Margaret Mead's investigation of adolescence in Samoa—was for a new morality, a democratic gospel, as a way to respond to the new technological potentialities. In keeping with William Ogburn's influential notion of the "cultural lag," the latter-day progressive intellectuals sought to reform social institutions to meet the challenges of modern technology. (See Jamison 1989.) In the arts and in functionalist architecture and design, this pragmatic position would become influential and would provide ideas and approaches that would enter into the "New Deal order" of the 1930s (Fraser and Gerstle 1989).

Thorstein Veblen and the Technocrats

C. Wright Mills once wrote that "there is no failure in American academic history quite so great as Veblen's." (See Veblen 1953: ix.) In the 1920s, however, Veblen had a brief period of glory as his writings gained him a certain following among technocratic progressives. In any event, his books from the early 1920s, *The Engineers and the Price System* (1921) and *Absentee Ownership and Business Enterprise in Recent Times: The Case of America* (1923), provided a view of technology's role in economic life that was influential. These later writings emphasized the central importance of the "instinct of workmanship" in industrial production; like a latter-day Marx, Veblen analyzed the transformation of the industrial order from one based on simple commodity production to a system based on the conscious use of technological research. Unlike Marx, however, Veblen had little faith in the working class to take over these new science-based means of production; instead, he urged the engineers to take greater responsibility for industrial development and management.

Whereas his prewar writings had been written from a professorial distance, in later life, having moved to New York and begun to write articles for a journal called *The Dial*, Veblen propounded a more explicitly political or ideological message. *The Dial* was a midwestern literary magazine that had been taken over by young progressives in 1916 and moved to New York, where Veblen joined John Dewey and the feminist writer Helen Marot in supporting the war effort and then in "the task of reconstructing society after World War I" (Perry 1989: 322). Unlike his more democratically inclined colleagues on the editorial board, Veblen used *The Dial* as a platform for technocratic elitism. His articles on the machine age and on the new role of engineers were later turned into books that were read by many corporate managers and engineers. His writings provided a kind of theoretical underpinning for the various efforts taken during the 1920s by politically and socially minded engineers, from scientific management to industrial research to political propaganda for "social engineering."

It is ironic that Veblen should be best known for inspiring an elitist movement; in actual fact, he was throughout his life an individualist and an outsider and about as far from power as anyone could be: that was his failure. His strength was in subjecting the new modes of production to systematic and critical scrutiny. Veblen's first book, *The Theory of the Leisure Class* (1899), had sought to identify the cultural constraints to an efficient industrial order in the "pecuniary" tastes of the businessmen and the nouveau riche. He was particularly critical of their ostentatious interest in rejecting machine-made products. As he put it, in his characteristically ironic way:

> . . . the generic feature of machine-made goods as compared with hand-wrought articles is their greater perfection in workmanship and greater accuracy in the detail execution of the design. Hence it comes about that the visible imperfections of the hand-wrought goods, being honorific, are accounted marks of superiority in point of beauty, or serviceability, or both. Hence has arisen that exaltation of the defective, of which John Ruskin and William Morris were such eager spokesmen in their time; and on this ground their propaganda of crudity and wasted effort has been taken up and carried forward since their time. And hence also the propaganda for a return to handicraft and household industry. (Veblen 1899: 115)

In this passage we can see how Veblen stated his arguments. Wit, exaggeration, and an unbridled belief in the progressive nature of "machine-made goods" are combined in order to question the competence of the business or "leisure" class. Veblen never swayed from his basic technocratic position, which was that the "machine age" required a new kind of planning and management and that those activities were best placed under the control of engineers. The captains of industry (Veblen's "leisure class"), having taken on many of the "archaic" or conservative trappings of the old aristocracy, served to constrain the logic of technology from effectively determining the rhythms and routines of life. Veblen was particularly scornful of his fellow academics, who had, he wrote, given "higher learning" a leisure-class orientation and thus made it largely useless.

In *The Theory of Business Enterprise* (1904) and *The Instinct of Workmanship and the State of the Industrial Arts* (1914), and in a number of articles published in economics journals, Veblen tried to develop a more "useful" kind of academic science. He tried to carve out a new economics that would focus on the processes of industrial change, as well as on the institutional aspects of economic life. His institutional economics was related to similar efforts elsewhere, especially that of Joseph Schumpeter in Austria, but Veblen's writings bore his own characteristic stamp.

Veblen's emphasis on the central importance of machinery in the industrial economy of the early twentieth century was not unique; many of his fellow "progressive" intellectuals—John Dewey, Charles Beard, Jane Addams, Randolph Bourne, Herbert Croly—recognized the new technical potentialities that had been unleashed by the linking of science with engineering, and by the infusion of systematic research into the production process. But no one singled technology out as the decisive force of social transformation to the extent that Veblen did. At the outset of his most ambitious philosophical work, *The Instinct of Workmanship*, he boldly stated:

It is assumed that in the growth of culture, as in its current maintenance, the facts of technological use and wont are fundamental and definitive, in the sense that they underlie and condition the scope and method of civilization in other than the technological respect, but not in such a sense as to preclude or overlook the degree in which these other conventions of any given civilization in their turn react on the state of the industrial arts. (Veblen 1914: 1)

In the book, Veblen sought to derive the technological urge not from cultural values and traditions or from economic or commercial necessity but from a psychological theory of instincts. In his wartime writings, he continued to read behavioral psychology in order to develop a more solid material grounding for his unwavering belief in modern technological development. His intriguing comparison of English and German industrialization in *Imperial Germany and the Industrial Revolution* (1915) was a first intervention into a more explicitly political discussion. The war, one might say, brought him out of the academic or theoretical phase of his career and into more "applied" concerns that would occupy him until his death in 1929. In *Imperial Germany*, he tried to explain the conflict between England and Germany as based on different modes of developing technology. The Great War, for Veblen, was a conflict between England and America's technological civilization and Germany's "Prussian-Imperial state." In Veblen's words: "The most characteristic habit of thought that pervades this modern civilization, in high or low degree, is what has, in the simplest terms hitherto given it, been called the mechanistic conception. Its practical working-out is the machine technology, of which the intellectual precipitate and counterpart is the exact sciences." (1915: 268–270). Veblen's argument was that the Prussian-Imperial system was a historical digression:

This warlike-dynastic diversion in which the Imperial State has been the protagonist is presumably of a transient nature, even though it can by no means be expected to be ephemeral. The Prussian-Imperial system may be taken as the type-form and embodiment of this reaction against the current of modern civilization. . . . In the long run, in point of the long-term habituation enforced by its discipline, the system is necessarily inimical to modern science and technology, as well as to the modern scheme of free or popular institutions, inasmuch as it is incompatible with the mechanistic animus that underlies these habits of thought. . . . (ibid.)

After the war, Veblen's irritation with the captains of industry and their pecuniary values grew in intensity. He argued, in a series of articles later published under the title *The Engineers and the Price System*, that the continued dominance of industrial production by financial managers rather than engineers constrained production. The captains of industry were not primarily interested in developing the industrial arts; governed by a commercial logic, they evaluated technology in terms of profitability and not necessarily in terms of productivity. In addition, the growing complexity and specialization of technological development created problems of understanding for the business leaders, who were "less and less capable of comprehending what is required in the ordinary way of industrial equipment and personnel" (Veblen 1921: 61–62).

For Veblen, it had become increasingly important that those who actually understood the machinery of industrial production be in charge of its development. The dead hand of finance was becoming a serious barrier to the future development of the industrial process. Inspired by the Bolshevik revolution in Russia, Veblen (1921: 163) called for a "soviet" of engineers: "The technicians may be said to represent the community at large in its industrial capacity, or in other words the industrial system as a going concern; whereas the business men speak for the commercial interest of the absentee owners, as a body which holds the industrial community in usufruct." The engineers needed to form a new political and cultural elite that could counter the unproductive pecuniary interests of the absentee owners. The development of technology required a cultural and political transformation if it was to go forward and not stagnate and ossify.

In his 1923 book *Absentee Ownership and Business Enterprise in Recent Times: The Case of America*, Veblen contrasted the habits and customs of American culture with the requirements of the "new order" of mechanization and technology-based industry. For Veblen the quintessential American value was trade and business, and throughout his life he castigated the businessmen and their vested interests for holding back the flow of mechanical innovations and technological progress. In *Absentee Ownership* he located the source of the problem in the dominance of the values of the country town: "The country town is one of the great American institutions, perhaps the greatest, in the sense that it has had and continues to have a greater part than any other in shaping public sentiment and giving character to American culture." (Veblen 1923: 142) For Veblen, the business of the country town was primarily real estate and retail trade, both of which promulgated an approach to success and economic activity that ran directly counter to the technological logic of modern industry. But now the future could no longer be held back: "The material conditions are progressively drawing together into such shape that this plain country-town common sense will no longer work." (ibid.: 165) Instead, Veblen and others called for an increased status and increased managerial role for engineers. Having proposed a Soviet of engineers and a revolution of technicians in 1919, Veblen now argued for a technocratic order. It was not sufficient for the engineers to become the managers of industry. There was also a need for the engineering mentality to be applied to the management of society itself. Without apparently calling them by name, Veblen (ibid: 273) argued for an increased responsibility for engineers in public administration and politics:

The safe and sane plan of common sense now dictates that industrial operations must be conducted by competent technicians. And this holds true in a special degree for the larger operations and the more formidable organizations of work and equipment, where many technological factors and a wide range of materials and processes are drawn together for teamwork in quantity production on an extensive scale. So it should also hold true in a superlative degree as regards the oversight and control of the industrial system at large as a going concern; the balance, articulation, and mutual support among the several lines of production and distribution that go to make up the system.

As Noble (1977) has shown, the 1920s were a time when many engineers did try to design corporate America in their image, and their achievements are still with us. This technocratic pole in the technology debate included many of the founders of management science, as well as the corporate research managers and production consultants. In the United States, the iconoclastic Veblen served as theorist, while Herbert Hoover came to embody the engineer as politician and state bureaucrat. In his activities at the Department of Commerce during the 1920s, Hoover sought to bring technological logic into the center of state economic policy.

The call for a "revolution" of the engineers was popular among certain intellectuals. There were Veblen clubs and study groups, and a number

of younger writers and even some engineers developed the ideas further
(Layton 1971; Tichi 1987). But Veblen himself slipped back into the
obscurity from which he had briefly emerged. He died in 1929, poor
and largely forgotten. The technocrats who in the 1930s claimed to take
their lead from his writings were never a very significant political force,
although they had some influence in some city and state governments.[1]
In the 1920s and later, Veblen's influence was actually based more on
his earlier writings—*The Theory of the Leisure Class* (1899) and *The Theory
of Business Enterprise* (1904)—and on his skeptical, even cynical attitude
toward the follies of his fellow Americans. His sharp analyses of the
technological age were sources of inspiration for many technocratically
oriented progressive intellectuals, from Walter Lippmann to Stuart
Chase. The position they developed in the 1920s was one of guarded
optimism—a kind of continuation, by experts, of the progressive reform
ambitions of the prewar years.

In *Tragedy of Waste* (1925) and *Men and Machines* (1929), Chase popu-
larized Veblen's basic arguments and outlined an approach to social
and political life that recognized the formative role of machinery. In his
commonsensical journalistic style, Chase translated Veblen's theories
into everyday language. He traced the history of the machine from
antiquity to the present, and he tried to distinguish the machine itself
from its myriad social consequences.

With a similar ambition but a more academic approach, the Harvard
economist Abbott Payson Usher produced a *History of Mechanical Inventions* (1929), in which he closely analyzed the ways in which machines
had actually been developed, using a psychological model of inventive
activity somewhat similar to Veblen's. (See Molella 1990.) The goal for
both the journalist Chase and the academic Usher (as it would be in
the 1930s for Schumpeter, who would join Usher at Harvard) was to
make the technological civilization function more efficiently in its own
terms, not by a revolution of engineers, but by a transfer of scientific
and technical rationality to economics and politics. Economists needed
to focus much more attention on the process of technological innovation,
Schumpeter argued. It was not the accumulation of capital or even
managerial skills that formed the basis for economic growth; it was rather
the fundamental process of "creative destruction" through which one
cluster of radical innovations replaced the older technologies that could

1. Some of the "system builders" of electricity distribution networks considered
themselves followers of Veblen (Hughes 1989).

no longer serve as motors for further growth and expansion. Chase and many of the later New Deal experts can be considered pioneers in technological assessment; they tried to evaluate particular technologies, to construct balance sheets of "pros and cons" that could be used by politicians and administrators in the choice of projects and techniques. Chase helped bring about the operationalization of Veblen's ideas that was to play a certain role in the social engineering that characterized at least part of the governmental activity of the 1930s (Pursell 1979). Schumpeter, on the other hand, after falling out of fashion in the 1950s, was himself rediscovered in the wake of economic decline in the 1970s, and has since inspired economists and engineers to continue to unravel the internal dynamics of technological innovation in hopes of disclosing a new "long wave" of expansion and growth (Jamison 1989).

The Traditionalist Critique of Joseph Wood Krutch

In later life, Joseph Wood Krutch wrote many books, primarily about the plants and animals in the deserts of the American Southwest, where he moved after World War II. In 1948 he wrote a biography of Thoreau. In the interwar years, Joseph Wood Krutch was a well-known drama critic for *The Nation* and a professor of English at Columbia University, where he received a doctorate in 1923 with a dissertation on "Comedy and Conscience in the Renaissance." Krutch, one might say, was a typical literary man, identifying with an earlier, more cultivated age, and seeing his and culture's role more generally in terms of moral or spiritual elevation. In his 1929 essay *The Modern Temper*, Krutch gave voice to a critical position that was not his alone, even though he gave it perhaps its most influential formulation. He was bringing the traditionalist position up to date, following in the footsteps of Henry Adams, who had berated the materialism of the age in his famous *Education*.

Science and technology had led modern man astray, Krutch contended. The material facts of the world, however true and reliable they might be, simply could not address the question of human existence. "We are disillusioned with the laboratory," he wrote, "not because we have lost faith in the truth of its findings but because we have lost faith in the power of those findings to help us as we had once hoped they might help." (Krutch 1929: 53) There had been a religious motivation to the scientific quest in the early modern period, but it had now withered in a spiritless drive for an ever more meaningless utility:

We went to science in search of light, not merely upon the nature of matter, but upon the nature óf man as well, and though that which we have received may be light of a sort, it is not adapted to our eyes and is not anything by which we can see. Since thought began we have groped in the dark among shadowy shapes, doubtfully aware of landmarks looming uncertainly here and there—of moral principles, human values, aims and ideals. We hoped for an illumination in which they would at last stand clearly and unmistakably forth, but instead they appear even less certain and less substantial than before—mere fancies and illusions generated by nerve actions that seem terribly remote from anything we can care about or based upon relativities that accident can shift. (ibid.: 47)

Krutch was totally uncompromising in his criticism, and *The Modern Temper* found a receptive audience among what might be termed the literary intelligentsia. As Pells (1973: 29) puts it: "The melancholy tone of its conclusions summarized the attitudes of a generation of writers weaned on Henry Adams, T. S. Eliot, and Oswald Spengler." For most other Americans, *The Modern Temper* was an unusually downbeat book, a depressed and highly skeptical view of the modern world. Science, for Krutch, had failed to deliver anything of importance to humanity, and now modern man was stuck with it whether he wanted it or not. Disillusion was the characteristic of the modern mood that Krutch identified: the dream of enlightenment through reason had been found to be ultimately meaningless.

For Krutch, modern society had rejected its humanity in pursuit of comfort, speed, business. But, he argued, "we cannot make physical speed an end to be pursued very long after we have discovered that it does not get us anywhere" (Krutch 1929: 56). In a final rhetorical flourish, Krutch made his choice clear: he would leave the future to "those who have faith in it." For his part, he would proclaim, in Shakespearean fashion:

Hail, horrors, hail,
Infernal world! and thou profoundest hell,
Receive thy new possessor.
(ibid.: 168)

Krutch would be on the side of man against nature and on that of culture against science. His was the first significant American anti-science tract of the twentieth century. By drawing the dichotomy as sharply as he did, he left no constructive role for culture to play in developing technology. Culture became the preservation of the non-technical, and eventually (as Krutch would write in the 1930s) literature was valuable to the extent that it did not try to be socially useful (Pells 1973: 184).

Krutch would inspire other literary figures to take up arms against the modern temper, but his would be a decidedly minority position in the interwar years. The more constructive approach of the "human ecologists" and the pragmatists would have a far greater influence on the range of activities that would take place during the New Deal era of the 1930s. But later Krutch had a change of heart and grew closer in perspective to Lewis Mumford, who had been so critical of *The Modern Temper*. In an interesting convergence, Krutch's more constructive ecological writings of the 1940s and the 1950s would complement Mumford's more disillusioned postwar writings, and both men would contribute to the making of an environmental consciousness amidst the countercultural enthusiasms of the 1960s (Fox 1985).

Lewis Mumford and the Human Ecological Critique of Technology

Lewis Mumford, born in 1895, came of age during World War I, and his thinking was strongly affected by the fading of the progressive dreams that came in the wake of the war. A man of many interests and talents, he was one of the most active "reinventors" of American culture in the 1920s, and in the 1930s he produced perhaps the most influential interpretation of the meanings of modern technology that any American intellectual has ever written. His views on technology evolved in the course of the interwar years, only to be shaken again by the use of the atomic bomb at the end of World War II. But that is another story.[2]

In his first book, *The Story of Utopias* (1922), Mumford sketched, in the words of Thomas (1990: 79), a "plan for the regional reconstruction of the United States which Mumford would develop, expand, modify, elaborate but never essentially change." Not yet 30 years old, Mumford was already an active public intellectual. Besides having written for the progressive journals *The New Republic* and *The Dial*, he was the secretary of the Regional Planning Association. Though he shared many of the prewar concerns of the progressive pragmatists, Mumford brought something new into the discursive framework: an ecological sensibility that he had adopted from the Scottish biologist and urban sociologist Patrick Geddes. For Mumford, as for Geddes, culture was primarily the geographical landscape—what we today would call the environment. After reviewing visions of the ideal society from Plato to William Morris, Mumford concluded not with a new utopian vision of his own but with a

2. See Mendelsohn 1990; see also Jamison and Eyerman 1994: 82–92.

proposal for a new kind of scientific practice—which, following Geddes, he called the Regional Survey. He admitted that science had "provided the factual data by means of which the industrialist, the inventor and the engineer have transformed the physical world; and without doubt the physical world has been transformed." "Unfortunately," he continued, "when science has furnished the data its work is at an end. . . . So far, science has not been used by people who regarded man and his institutions scientifically. The application of the scientific method to man and his institutions has hardly been attempted." (Mumford 1922: 271–272) Mumford went on, as he would continue to do throughout the interwar period, not to criticize science, but to suggest ways to complement its cold truths with a sense of life. On the one hand, he argued that the specialized knowledge of the scientist should be placed within a more general viewpoint, and that the factual orientation of the scientist should be balanced with the emotional and subjective wisdom of the artist. Most importantly, the abstractions of science should be connected to real life. The danger in separating thought from action was that scientists lost a sense of proportion or limits in extrapolating from their own limited experience, and the general public tended to lose contact with the actual practice of science. "The upshot of this dissociation of science and social life is that superstition takes the place of science among the common run of men, as a more easily apprehended version of reality" (ibid.: 275). The Regional Survey, as Mumford outlined it, was a way to cultivate a more socially useful and relevant technological development:

The aim of the Regional Survey is to take a geographic region and explore it in every aspect. It differs from the social survey with which we are acquainted in America in that it is not chiefly a survey of evils; it is, rather, a survey of the existing conditions in all their aspects; and it emphasizes to a much greater extent than the social survey the natural characteristics of the environment, as they are discovered by the geologist, the zoologist, the ecologist—in addition to the development of natural and human conditions in the historic past, as presented by the anthropologist, the archeologist, and the historian. In short, the regional survey attempts a local synthesis of all the specialist "knowledges." (ibid.: 279)

Mumford thus developed an interdisciplinary ideal of science that combined ecology, history, geography, and sociology. He emphasized the region, or the local context, as the basis for all development. As he explored American history over the next few years in search of a "usable past," and as he worked as secretary of the Regional Planning Association, he would continue to develop what might be termed a human ecological critique of modern American civilization. Mumford's perspective resembled that of other groups of human ecologists which emerged in the interwar years, such as the "southern regionalism" developed by the sociologist Howard Odum at the University of North Carolina and the urban ecology developed by Robert Park and his colleagues in Chicago.

Mumford's most influential publication of the 1920s was *The Golden Day* (1926), which George Santayana, in a jacket endorsement, called "the best book about America, if not the best American book that I have ever read." By 1926, something of an intellectual movement to reevaluate the artistic achievements of nineteenth-century America had developed among literary and artistic critics. This movement, begun before the war in the writings of Waldo Frank and Van Wyck Brooks, continued with the *American Caravan* yearbooks and a number of other works. Mumford's book was distinguished from those of the literati by his ecological view of culture and by his active use of history to evaluate contemporary writers and standpoints. Even though much in *The Golden Day* was new and interesting, it was Mumford's critique of the pragmatism of William James and John Dewey that apparently attracted the most notoriety at the time. Mumford took issue not with technology itself but with the materialism it had inspired, both in philosophy and in life, and he saw that materialism as being most clearly formulated in the writings of Dewey.

In *The Golden Day*, Mumford sought to affirm the significance of the classic writers of the nineteenth century as the representatives of a particular American mentality or identity. Hawthorne, Emerson, Thoreau, and Whitman, he claimed, had articulated a different kind of modern sensibility, framed by the experience of the natural environment and by the "republican" values of those who had tried to tame it. It was the meeting of a cultivated, European mindset with the raw American landscape that had formed the basis for a distinct national art and literature. The writers whom Mumford would continue to praise throughout his long life had brought on the morning of the golden day; by the end of the century, with the closing of the frontier, night had come, and the engineer had become the cultural hero: "The Edisons and Carnegies came to take the place in the popular imagination once occupied by Davy Crockett and Buffalo Bill." (Mumford 1926: 118) As Tichi (1987) has shown, a number of popular novels appeared at that time celebrating the work of the engineer as the modern-day embodiment of American virtue. This was also the time when some engineers had become industrialists, or at least had joined forces with industrialists to create new forms of economic and technological development. It was not by chance that Edison and Carnegie—an inventor and a corporate leader—were linked together in Mumford's mind. For Mumford, the glorification of engineering was part of a broader transformation of American culture in the direction of consolidation, regimentation, and order: "The great captains of industry controlled the fabrication of profits with a military discipline: they waged campaigns against their competitors which needed only the actual instruments of warfare to equal that art in ruthlessness." (Mumford 1926: 119)

Mumford criticized Dewey, James, and other writers of the progressive era for their "acquiescence"—their acceptance of the new technological society. In trying to control its excesses, he argued, they succumbed to its materialist values. Much as the realist novelists of the era—Theodore Dreiser, Frank Norris, Upton Sinclair—could describe the problems they saw but could not envision an alternative social order, Dewey's pragmatic philosophy was, according to Mumford, a product of the time and place in which it was formulated: "No one has plumbed the bottom of Mr. Dewey's philosophy who does not feel in back of it the shapelessness, the faith in the current go of things, and general utilitarian idealism of Chicago." (ibid.: 131) Mumford had taken Randolph Bourne's critique of Dewey as his own. Pragmatism, and progressive thought in general, represented a failure of imagination; the glorification of the machine had eliminated the utopian side of man, and with it the possibility of a cultural evaluation of the contemporary technological civilization:

As Bourne said, the whole industrial world—and instrumentalism is only its highest conscious expression—has taken values for granted; and the result is that we are the victims of any chance set of values which happens to be left over from the past, or to become the fashion. An instrumental philosophy which was oriented toward a whole life would begin . . . by a criticism of this one-sided idealization of practical contrivance. . . . Without vision, the pragmatists perish. (ibid.: 137–138)

There were, and there would continue to be, important differences between Mumford and Dewey, but Mumford was not always justified in his critique of Dewey. Both a difference in generational sensibility and a difference in standpoint and intellectual orientation seem to have been at work (Westbrook 1991: 380ff.). Mumford, coming of age in the 1920s, saw Dewey (and Veblen, for that matter) as part of the problem. His criticism of pragmatism was a critique of the older generation. But it was also a critique of an overly positive or adaptive attitude to technology. Mumford, for all his utopian envisioning of an alternative technological order, remained throughout his life a critic of the values he saw as intrinsic to modern technological civilization. If he pointed to some positive potentials in technology in his writings of the 1930s—the writings that brought his position closer to Dewey's—he nonetheless remained a cultural critic of technology. The same cannot really be said of Dewey, and certainly not of Veblen. In the course of the 1920s, however, the two elder statesmen of progressive thought responded rather differently to the increasingly technocratic climate of the times. While Veblen became a kind of prophet for the young turks of technocracy, Dewey rethought the relationship between democracy and philosophy. Both men remained, as it were, faithful to the power of technology, but they looked to different actors and developed different political strategies for the necessary social and economic renewal.

Mumford, for his part, delved deeper into the meanings of the machine. As the 1920s wore on, he grew more interested in the creative potentialities of the new science-based technologies. One can see in his writings of the 1930s—particularly in *Technics and Civilization*, which he published in 1934—a far more sympathetic view of the industrial civilization than he had expressed in the 1920s. In the wake of the Depression, he realized that technology's promise was still largely unfulfilled and its human potential largely unexplored. The task of a cultural critic was not merely to identify problems but to provide constructive ideas for bringing the machine under human control, and, even more, using machinery to enrich human life. It had become clear that the machine needed a conscious program for its guidance; and it was such a program that Mumford aimed to provide, however contradictory it eventually turned out to be (Blake 1990: 279ff.).

Miller (1989: 300) points to the significance of Oswald Spengler's *Decline of the West* in the development of Mumford's thinking: "That big and oddly brilliant book, the work of a reclusive German schoolteacher, had been on Mumford's mind since 1926, when he reviewed the English translation of the first volume for *The New Republic*." Spengler offered to Mumford a view of history as moral prophecy, as well as an organic way of thinking about the development of civilization that well fit Mumford's own ideas. According to Miller (ibid.: 302), "Mumford agreed with Spengler that Faustian culture had entered the 'winter' of its development; but where Spengler peered into the future and saw only spreading blackness and blight, Mumford saw a brilliant post-Faustian world, a great revival of the regional and organic outlook.''

It is important not to exaggerate the similarity between Spengler and Mumford, however. Though they shared an organic approach to human history, Mumford was perhaps somewhat more sympathetic to modern science and technology and less opposed to the tenor of modern civilization than Spengler and other "traditionalist" critics. When he reviewed Krutch's *Modern Temper* in 1930, he explicitly renounced the attitude of despair that he saw in it. "We would not destroy the rigorous method of science or the resourceful technology of the engineer," Mumford wrote in his review (quoted on p. 31 of Pells 1973). "We would merely limit their application to intelligible and humane purposes. Nor would we remove altogether the mechanical world-picture, with its austere symbolism; we would rather expand it and supplement it with a vision of life which drew upon other needs of the personality than the crude will-to-power.''

In 1930, when asked to give a course at Columbia University on the Machine Age in America, Mumford made it the occasion for a much more ambitious project than any he had previously carried out. He read widely in the history of technology, and he won a fellowship to visit the technical museums in Europe in order to prepare himself for what he increasingly saw as his great book, his new synthesis. *Technics and Civilization* would actually be the first in a series of four books, to which Mumford, in his typically immodest way, would give the overall title "The Renewal of Life." The first book, in any case, was an original and exceedingly stimulating, if highly personal, reflection on the history of technology that argued for a new role for culture and for cultural analysis in the social process of appropriation.

In *Technics and Civilization*, Mumford applied his organic philosophy of history to technology, drawing on both his ecological perspective and his humanistic background. He described the rise of the machine as a cyclical process, and showed how the development of technology was itself a product of culture—and of cultural criticism. The machine civilization, he observed, had been prepared through centuries of institutional and intellectual developments. He noted that the ideas of mechanization and work discipline and the mechanical clock (which he considered the key invention) had emerged from monasteries and among theologians during the Middle Ages. He recalled the artist-engineers of the Renaissance, with their attention to the details of nature and their superhuman ambitions to encompass all knowledge and revolutionize all practical skills. He discussed the metaphysics of the machine, the quantification of time and human relations that had accompanied the rise of capitalism, and the regimentation of space and environment that had come with the great transformations of the medieval landscape. And, in his colorful and all-encompassing way, he disclosed the cultural preconditions for the development of the modern technological universe. Technics, Mumford argued, were not merely the "productive forces" that Marx and latter-day economists had focused their analytical gaze upon. They were, rather, manifestations of the entire range of human attributes and motivations, from playfulness to aggression and from ingenuity to artistry. It was the humanness of technology, in all its complexity and creative chaos, that Mumford sought to articulate in his remarkable book.

But Mumford also tried to bring some order to the chaos by distinguishing the various waves of mechanization from one another. For this purpose, he borrowed from Patrick Geddes a terminology and a historical framework which Geddes had adapted from archaeology (Williams 1990). First had come the *eotechnic* wave, based on water power and primarily using wood as the working material, when technics were well integrated into the surrounding landscape; then had come the *paleotechnic* nightmare of the industrial revolution, when coal and iron had brought about a totally different and far less attractive technical regime. In the twentieth century, a third period, a *neotechnic* epoch, could be distinguished, in which technological innovation was based largely on applied science, and a new organic guiding principle could be discerned in relation to technical development:

> ... we have now reached a point in the development of technology itself where the organic has begun to dominate the machine. Instead of simplifying the organic, to make it intelligibly mechanical, as was necessary for the great eotechnic and paleotechnic inventions, we have begun to complicate the mechanical, in order to make it more organic: therefore more effective, more harmonious with our living environment. ... One can now say definitely, as one could not fifty years ago, that there is a fresh gathering of forces on the side of life. The claims of life, once expressed solely by the Romantics and by the more archaic social groups and institutions of society, are now beginning to be represented at the very heart of technics itself. (Mumford 1934: 367, 368)

Mumford aimed to humanize technology—to link its development to broader cultural patterns and movements—but also to give the machine, as it were, a life of its own—a life cycle, from the infancy of the Middle Ages, through the wild aggressive youth of the nineteenth century, to the potential maturity of the twentieth century. In both senses, Mumford wanted to breathe life into what was typically seen as cold, hard machinery. He saw in many of the new science-based technologies a biological or organic essence that was superseding the mechanical philosophy of the eighteenth and nineteenth centuries. These technologies expanded man's senses and powers rather than replacing them or reducing them to the merely mechanical. He also saw opportunities for assimilating the machine into patterns of regional organization—opportunities he would discuss in more detail in his 1938 book *The Culture of Cities*. The great promise of the science-based technologies was that they were amenable to decentralization and to democratic control.

Perhaps most significantly, Mumford saw a new aesthetics emerging— a kind of art not only embodied in the technological products but also made possible by the new instruments of artistic reproduction. For Mumford, photography, recorded music, and moving pictures expanded the human personality: "Whereas in industry the machine may properly replace the human being when he has been reduced to an automaton, in the arts the machine can only extend and deepen man's original functions and intuitions" (Mumford 1934: 343). Modern man expressed himself by means of technological instruments. He could record his feelings and portray his environment in ways that enhanced his experience of life, and that, for Mumford, was always the main goal of all activity.

Mumford's contribution was to provide a personal (some might say overly personal) reading of the multiple meanings of the mechanical order. Through his organic philosophy, which was at once humanist and ecological, Mumford could characterize the social and human components of technology in a more comprehensive way than any other author had. *Technics and Civilization* created a new field of study: history of technology. Although many would follow Mumford in subjecting technology to historical analysis, particularly in the postwar United States, no one, not even Mumford himself, would manage to take so much into account as he did in that book. Though its flowery and opinionated language may sometimes irritate and its lack of references may annoy the academic reader, the sheer amount of thought continues to fascinate. Mumford succeeded in placing technological development in a human context. After *Technics and Civilization*, the debate about technology moved to a new level of constructive ambition and seriousness. It would never be quite the same.

John Dewey and the Pragmatic Assessment of Technology

John Dewey, as has already been noted, was one of the main articulators of the progressive standpoint at the turn of the twentieth century. Even more than William James, Dewey had sought to reconstruct philosophy along modern, experimental lines. Dewey had also sought to link philosophy with progressive practice, particularly in his pedagogical activities and in his work at the University of Chicago's Laboratory School.

By the 1920s, Dewey had grown critical with some of his earlier positions. He had come to realize that the war—and his support for the United States' involvement in it—had been a mistake, a failure of democratic principles. The war had not ushered in a more equitable world order, as Woodrow Wilson had promised; rather, there had emerged a new era of imperialism and global competition. Particularly important in Dewey's later development was his confrontation with another civilization when he visited China in 1919 in the midst of the May 4th Movement (the student revolt that would lead to the republic of Sun Yat-Sen). Also important were the criticisms of his ideas that had been expressed during the war by Randolph Bourne. For Dewey the 1920s were a period of reassessment that culminated in a series of writings in which he directly addressed the concerns of Mumford and Krutch. For our purposes, the key text is *The Quest for Certainty* (published in 1929, the same year as Krutch's book *The Modern Temper*), in which Dewey reaffirmed his pragmatic faith but also offered perhaps his most reflective and searching defense of science and technology.

What was important for Dewey was to distinguish scientific facts from reality. Westbrook (1991: 356) quotes him as follows: "The man who is disappointed and tragic because he cannot wear a loom is in reality no more ridiculous than are the persons who are troubled because the objects of scientific conception of natural things have not the same uses and values as the things of direct experience." Dewey was critical of Krutch and other anti-modernists because they misunderstood what science actually produced, which, he argued, was not a knowledge of things but an instrumentally mediated knowledge of relations: "The relations a thing sustains are hardly a competitor to the thing itself." (ibid.: 355) It was also important to distinguish between scientific knowledge and the positivist philosophy of science, and, as always, Dewey opposed those philosophers who valued reason and logic above practical judgment. Instead of reason, Dewey argued for intelligence in public affairs and in the cultural control of science and technology:

A man is intelligent not in virtue of having reason which grasps first and indemonstrable truths about fixed principles, in order to reason deductively from them to the particulars which they govern, but in virtue of his capacity to estimate the possibilities of a situation and to act in accordance with his estimate. In the large sense of the term, intelligence is as practical as reason is theoretical. (ibid.: 357)

The point for Dewey, as an egalitarian, was to keep science and scientific reason firmly in their place, under democratic control. In the words of Westbrook (ibid.: 360), "For Dewey, substituting intelligence for transcendent reason . . . in the moral life of a culture required vesting authority not in the insights of philosophers (or scientists) but in the deliberations of ordinary men and women skilled in the art of practical judgment." Dewey was, above all else, a philosopher of democracy who came to see that technology was becoming as much a threat as a contributor to American democratic behavior. What Dewey promulgated was a democratic technics, a functional and egalitarian version of utilitarianism, and as he continued to "reconstruct" his philosophy in the interwar years and to serve as a spokesman for a new wave of pragmatic populism among intellectuals he developed a more skeptical view of technology (Hickman 1990). He no longer identified science and technology with progress, and he no longer saw the linking of technology and democracy as inevitable. Ever more radically, in the 1930s he called for a new morality to deal with the consequences of science and technology. He put it this way on page 154 of *Freedom and Culture*, a widely read book published in 1939:

Science through its physical technological consequences is now determining the relations which human beings, severally and in groups, sustain to one another. If it is incapable of developing moral techniques which will also determine these relations, the split in modern culture goes so deep that not only democracy but all civilized values are doomed. . . . A culture which permits science to destroy traditional values but which distrusts its power to create new ones is a culture which is destroying itself. War is a symptom as well as a cause of the inner division.

By 1939, Dewey was already 79 years old and long retired. As an elderly philosopher, he certainly had some significance in the interwar technology debates, but it was perhaps more through his disciples that his position was articulated and further disseminated. In the 1920s, a group of progressive social scientists, following in the footsteps of Dewey's philosophy if not adhering to the letter, had developed a form of pragmatic sociology that would play an important part in the practical efforts in the interwar years to "assess" the social consequences of technology. Robert Lynd, William Ogburn, and many other social scientists would "reinvent" pragmatism as a kind of social engineering. (See Ross 1991.) And they would not be without influence in shaping the New Deal. Indeed, many of the reform activities of the 1930s were conscious attempts to bring together the various positions that had been staked out in the technology debates of the 1920s. The task in the 1930s was increasingly seen as one of creating a new national project that could combine the best of the assimilators and the adapters, the elitists and the egalitarians. In the words of Susman (1984: 156): "A key structural element in a historical reconstruction of the 1930s is the effort to find, characterize and adapt to an American way of life as distinguished from the material achievements (and the failures) of an American industrial civilization."

Conclusions

The four positions that have been discussed in this chapter were certainly not the only ones that were articulated in the interwar years. But they do seem to represent the poles of a wide-ranging debate about the meanings of modern technology that continued into the postwar era, and even into the present. Out of these four positions have grown several distinctive "discourses," which have even manifested themselves in the formation of academic specialties.

On the one side, there have been those who have identified America's greatness and global responsibility with technological superiority. That position would grow in strength with the development of the atomic bomb and the more general experience with "big science" after World War II. In the 1950s, technocracy would gain a new lease on life through the largesse and the institutionalization of what came to be called the military-industrial complex. Latter-day followers of Thorstein Veblen, such as the economist John Kenneth Galbraith, would see in the "techno-structure" of American corporations a new source of economic dynamism and social transformation as well as a new form of political power.

And for the small but growing group of institutional or innovation economists, Veblen is seen as a kind of founding father, even though his own views have tended to be neglected as the years have passed.

On the other side, among the humanists, the elderly Lewis Mumford himself would be one of the most outspoken critics of the megamachine that the United States had become in his eyes. In the postwar era, Mumford's writings would help shape a new range of humanist disciplines, from the history of technology to human ecology, that would explore ways to assess modern technology's impact on the natural environment as well as its impact on social relations (see Jamison and Eyerman 1994). Joseph Wood Krutch would be one of several intellectuals who would carry the criticism of modernity further into an appreciation of the ways of life of the traditional cultures of North America. The Indian spirit that Krutch wrote about from his new home in the American Southwest would be a source of the countercultural movements that developed in the 1960s, which have continued to inspire environmentalism and broader cultural practices. The pragmatism of John Dewey, downplayed and largely forgotten in the years immediately after World War II, has been rediscovered as an indigenous American kind of philosophy, at once practical, democratic, and sophisticated. In the writings of Richard Rorty and other philosophers, Dewey's instrumentalism has come alive again to contribute to contemporary debates about technology, science, and the kinds of knowledge that are appropriate for a post-industrial age.

The positions that were staked out in the interwar years have thus continued to inform the intellectual appropriation of technology in the United States. As engineers and corporate planners fashion new forms of "virtual reality" and as environmental extremists oppose further encroachments into the wilderness, there may be something to learn from the debates of the interwar years, when the terms of reference used in many of today's disputes were first established. The battle between the technocrats and the humanists goes on.

5

Engineering Cultures: European Appropriations of Americanism
Kjetil Jakobsen, Ketil G. Andersen, Tor Halvorsen, and Sissel Myklebust

The Brave New World

Ever since Benjamin Franklin achieved celebrity in Parisian society in the 1780s, European intellectuals have sensed that the United States was the land of the future. In 1830, when the aging Hegel last gave his introductory lectures on the philosophy of history, he noted that the future belonged not to Germany but to the United States. Hegel, however, refrained from further comments, his subject being history and America being precisely "a land of desire for people who have grown bored with Europe's rusty armory of history" (Hegel 1991: 209). What European intellectuals tended to see in the United States in the 100 years that followed the War of Independence was the coming of democracy. In 1792 the Jacobins enthusiastically set up their national convention on the model of the Philadelphia convention. Fifty years later, Alexis de Tocqueville wrote his lucidly critical books *De la Democratie en Amerique.*

For Europeans the "problem of America" came to a head in the early decades of the twentieth century, by which time "Americanism" had taken on new meanings. The United States was no longer seen as the land of democracy; it was now the land of science-and-technology-generated affluence. American society represented new relationships among science, technology, and culture. By the 1950s these relationships would almost define the twentieth-century discourse on modernity, but in the interwar years Americanism met some fierce opposition and was reconstructed according to the meaning horizons and practical interests of social groups in the various European nations.

The appropriation of Americanism was never politically neutral. In his essay "Between Taylorism and Technocracy," Charles Maier (1970) notes that American methods of production were embraced as the gospel of the future by revolutionaries (of both the extreme right and the extreme left: the futurists and fascists of Italy and the Bolsheviks of Russia) but also, somewhat paradoxically, by Gustave Herriot in France and Walther Rathenau in Germany—social reformers who were searching for a middle way that would eliminate class warfare. Anti-Americanism was voiced by conservative humanists, by laissez-faire liberals, and by traditional social democrats.

National trajectories may be distinguished not only in regard to whether Americanism was to be seen as a good or a bad thing but also, and more importantly, in regard to exactly what it meant. In Germany, France, and England, "boosters" and "knockers" applied apparently similar arguments for and against Americanism but were actually, more often than not, discussing different things, since the terms of the debate—"Americanism," "Fordism," "Taylorism," "rationalization," "science," "technology," and so on—tended to mean different things in the various countries.

We start from a comparison of national engineering styles—American, English, French, and German—at the turn of the twentieth century. The national models are ideal types, and we emphasize differences rather than similarities. In the tradition of Max Weber, ideal types are not hypotheses but historical concepts that may be helpful in reconstructing the dynamics of past developments. The institutions that shape technological discourse—educational systems, research councils, and professional organizations, but also languages and political structures—tend to work on national levels. We seek to demonstrate how, in the first half of the twentieth century, national differences in how "technology" had been situated within the nationwide systems of knowledge production gave rise to nationally specific intellectual discourses on "the technology question"—that is, specific ways of appropriating Americanism.

It seems paradoxical that, whereas in the nineteenth century the United States had been the most important source of inspiration for European democracy, in the twentieth century it became the main source of inspiration for European technocracy. Many European intellectuals, especially in Germany, tended to associate technocracy with democracy, seeing both as symptoms of the decay of culture in "the age of the masses."

The Engineers

The classic studies on engineers agree on the contradictory nature of their role. In *The Engineers and the Price System,* Thorstein Veblen described how the technologists' knowledge of the productive potential of the industrial system clashed with the interest of the "captains of industry," who "sabotaged" production in order to keep prices up. The situation, Veblen (1921: 166) claimed, was in principle ready for a "self-selected, but inclusive, Soviet of technicians to take over the economic affairs of the country and to allow or disallow what they may agree on, provided always that they live within the requirements of that state of the industrial arts whose keepers they are."

No such Soviet of engineers has appeared. In a widely acclaimed study of the American engineers, Edwin Layton (1971) suggests that Veblen misrepresented the basic dilemma of the engineer. What troubles the engineers about modernity is not capitalism but bureaucracy, says Layton. The engineer wishes to be a free professional, but is by necessity caught in the bureaucracy of the corporation. Layton's historical analysis is no doubt more adequate than Veblen's. However, Layton is writing exclusively about the American engineer, and his analysis is based on the Anglo-Saxon notion of the free professional. Layton's ideal types are therefore less useful for analyzing engineering professions in non-Anglo-Saxon countries. Our comparative analysis utilizes a model that differs both from Veblen's dichotomy between production and business and from Layton's dichotomy between professionalism and bureaucracy, but which encompasses certain aspects of both.

In the engineers' mode of legitimization there is a fundamental duality. The historical role of the engineer is the production of tools that serve to overcome the shortage of goods. But modernity is characterized by the differentiation of the premodern "household"—the *oikos*—into household and firm. The activity of the capitalist firm is not in any immediate way directed toward overcoming the shortage of goods. The minimum factor in capitalist activity is the shortage of capital. Shortage is not connected to the ability to reap enough goods from nature to sustain consumption. Under capitalism, production remains socially meaningless unless capital is mobilized in such a way that production can be realized on the market. This takes managerial knowledge. Management cannot simply be written off as "sabotage." Organizational knowledge forms an irreducible part of engineering work, and would do so in any highly differentiated economy. The managerial role of an engineer is a necessary supplement to the historical role. The ethos of production and the ethos of management may contradict each other in certain situations. As a salaried official in a capitalist corporation, an engineer may find himself occupied more with shutting down production than with producing.

Another dilemma related to engineering has had to do with how the knowledge practices of the historical role are to be represented. The engineer has had to choose between conceptualizing engineering activity as a practical art and as applied science. That there is a real choice here is often forgotten. Both natural scientists and engineers have at times had a professional interest in overstating the role of science in technological development—the scientists for funding reasons, the engineers in order to strengthen their professional bargaining power versus skilled workers and business executives. Notions like "applied science," "science-based industry," and "scientific management" borrow persuasive power from an intellectualized model of work derived from Aristotle. Work is seen as divided into conception and execution (*noiesis* and *poiesis*), with theoretical conception presumably preceding practical execution.

National Trajectories

The advent of Americanism marked a discontinuity in the European process of modernization. To keen observers before World War I, and to everyone after the war, the United States was the wealthiest and most powerful country in the world. Americanism, therefore, could not be dismissed easily. With Taylorism, Fordism, standardization, interchangeable parts, mass consumption, mass entertainment, the scientific professional, the social sciences, and the giant corporation, Americanism, perhaps, deserved to be called a new project of modernity. Generally speaking, the French intellectual establishment responded favorably to this new modernity, while reactions in Germany and England were mixed or negative.

Why was Americanism appropriated so differently in the three countries? For analytical purposes, the intellectual appropriators of technology and science can be divided into two groups: engineers and natural scientists presenting their knowledge practices to the public, and literary intellectuals writing on technology and science. In what follows, we will concentrate on the technologists.

A review of how engineering discourses were constructed historically shows that the greatest contrast was that between Britain and the continental countries, the professional organizations having been paramount in Britain and the state on the continent. More specifically, whereas the French style of engineering was shaped by the relationship between the state and the education system, British nineteenth-century engineering was shaped through interaction between industry and the professional associations, the state and the education system being relatively unimportant. In Germany the relationship between industry and education was decisive (Halvorsen 1993).

In the United States and Britain, technological occupations constructed themselves "autonomously" around the notion of "the professional," which in the twentieth century was reconstructed as the scientific professional. In France and the German states, academic engineering education was established about 100 years earlier than in the Anglo-Saxon countries. The occupation of engineer was constructed by the state as part of an explicit strategy of modernization. In twentieth-century England and America, engineering and natural science education tended to merge. In nineteenth-century France and Germany, they were institutionally separated. Academic engineers were educated in technological universities, natural scientists in "ordinary" universities. Unable to borrow persuasive power from the state, twentieth-century American and English engineers entered instead into a dependent relationship with natural science, drawing professional power from the notion of "applied science." The French and German engineers had less autonomy in regard to the state, more in regard to natural science.

The main difference in content between the Anglo-Saxon and the continental engineering discourses is neatly reflected in terminology. French and German language distinguish between "technology" and "technique." The distinction is analogous to the distinction between "epistemology" and "knowledge." "Technology" is the science of "technique." The Anglo-Saxon "applied science" discourse had no need for such a distinction. The science of technique was science. The technology-technique distinction therefore disappeared from the English language—an instructive example of how language and thought are shaped by academic institutionalization.

In the United States the concept of "applied science" was transferred from one field to another as part of the spread of professionalism in "the progressive era." Medicine was the pioneer in denoting its practices "applied science," but engineering followed closely (Wiebe 1967; Kimball 1992). In engineering, applied sciences triumphed first within the practices of the historical role, then, with Taylorism, within those of the managerial role.

A main reason why the British and American engineering fields differed is that, in their struggle for professional legitimation, the British and American engineers competed with different knowledge practices.

The alliance with science came late in England. "Engineer" was a loose category that covered all sorts of practical men, from great entrepreneurs to humble technicians. "Rule of thumb" had served British industry well in the nineteenth century and was hard to unseat. Starting with the establishment of the Imperial college in 1907, serious efforts were made to overcome a perceived lack of scientific competence in British industry. These efforts were, however, subject to what might be termed the academic bias of British culture. To be attractive on the job market, the engineers had to comply with the gentlemanly lifestyle ideals of the hegemonic Oxbridge liberal education. The English academic engineers therefore generally chose to call themselves scientists rather than engineers. British twentieth-century engineering education became grossly dualistic. Practical technological work tended to remain in the hands of engineers with little formal training, while a growing mass of overtly theoretically educated "applied scientists" struggled to unseat the humanistically educated Oxbridge men as candidates for prestigious positions in British industry.

The American engineering profession was somewhat more homogeneous. Engineering and "applied science" melted together in a common discourse. The engineers borrowed persuasive power from the rhetoric of natural science, but without giving up their identity as engineers. Engineering professionalism did not primarily develop in rivalry with the liberal arts. During the progressive era, American educational institutions introduced the social sciences (Ross 1991). Marginalist economics, sociology, political science, anthropology, and psychology developed perspectives that would, if they had not been answered, have questioned the managerial competence of the American engineer. The result was Taylorism.

Scientific Management

Despite the fact that it has been widely criticized, condemned, and ridiculed on both sides of the Atlantic for almost a century, there is a widespread feeling among historians and others that scientific management gives a clue to long-term developments in twentieth-century history. Three lines of argument may be distinguished.

The oldest line claims that scientific management "solved" the problem of class in capitalism. A recent and influential formulation is found within the French school of "Regulation."[1] This school claims that the scientific management of Frederick Winslow Taylor, along with Henry Ford's assembly line and "five-dollar day," constituted a twentieth-century "mode of accumulation" that was fundamentally different from nineteenth-century liberal capitalism. On the one hand, Taylorism and Fordism induced a class compromise at the level of production. Workers gave up their skills and their control over the labor process in exchange for steadily rising real wages. On the other hand, the increasing scale of production and the dependency on a domestic market, typical of this mode of accumulation, created a functional need for Keynesian and corporatist political regulation of the economy. The industrial practices of Taylor and Ford functionally induced the modes of regulation typical of capitalism during the period of rapid economic growth in the 1950s and the 1960s. The argument presupposes that Taylorism and Fordism were efficient management strategies that actually raised productivity above what was possible with other strategies. In the 1980s, the relative stagnation of the American economy relative to the Japanese and German economies led a number of writers to question this assumption. (See, e.g., Piore and Sabel 1984.)

A second line of argument claims that scientific management was important, not as a factor in production, but as a factor in the formation of ideology. Taylorism helped make the new professional middle class politically powerful. Faced with the late-nineteenth-century stalemate between the economic capital of the bourgeoisie and the political capital of the working class, the new professional middle classes used their cultural capital, based on science and technology, to construct a discourse that spoke of modernizing society through the "visible hand" of science, replacing class identity by meritocracy. This discourse of modernization was to gain hegemony in twentieth-century politics. Taylorism played a crucial part in its construction by addressing the problems in the key arena of the factory, establishing a privileged "scientific" point of view from which both labor and capital could be criticized.[2]

1. The "founding text" of the school of Regulation is Michel Aglietta's *Theory of Capitalist Regulation* (1979). Aglietta attempted nothing less than to write a new and updated version of Marx' *Capital* based on the US twentieth-century experience rather than on the British nineteenth-century experience. Other well-known contributors to the school are Alain Lipietz and Robert Boyer. Antonio Gramsci's writings are an important source of inspiration.

2. Merkle (1980) sees Taylorism as codified middle-class ideology. Haber (1964) emphasizes the importance of Taylorism in early-twentieth-century reform.

The above two lines of argument might be termed the Marxist line and the neo-Marxist line. Without neglecting these aspects, we would argue that the importance of scientific management is best grasped from a third angle. Drawing on Foucaultian themes, one could say that scientific management appropriated industrial organization within a new discourse. From the previously discrete discourses of science, technology, and organization, Taylor abstracted an "essence," thus unifying them into an epistemologically continuous discourse. Since the appearance of scientific management, it has been common for people to think of science, technology, and industry as being ruled by a single rationality, which is at the same time "instrumental" and scientific.

Within political life, Taylorism probably had its most profound and direct impact on communism. The enthusiasm of Lenin, Trotsky, and Stalin for scientific management is well known. Industrial labor under Soviet socialism was a far cry from the young Karl Marx's ideals of humanizing work, and scientific management had something to do with that. However, the social constructionist approach to Taylorism suggests that its prime impact is not to be found on the factory floor. What is at stake here is really the whole concept of the planned economy. Marx analyzed capitalism in great detail, but said little about socialism. In the 1890s the aging Friedrich Engels and the young Karl Kautsky began to speak about the emerging giant corporation as a possible model for socialism. The concept of the planned economy was, however, not worked out until World War I. The inspiration came from the wartime economy and from scientific management.[3] Wickard von Moellendorf, who coined the word *Planwirtschaft* (planned economy) in 1917, was a Taylorist engineer and a high ranking wartime planner (Hardach 1977: 58). What scientific management added to socialism was the vision of scientific planning. The rational unity of organization, technology, and

3. In the Taylor system, implementing time studies required fundamental changes in industrial organization. Choosing day-to-day tasks and preparing the written instructions called for a "planning room." The sequences in which machines and materials were to be used had to be specified, and the standardization of machines, tools, and operations had to be completed or the specified time schedules would be meaningless. In addition, the activities of a large number of workers were to be coordinated—a real challenge in view of Taylor's insistence on a far-reaching division of labor, with tasks being adjusted to individual training. To compile all these efforts, it was necessary to introduce systematic planning, and the planning department became the nerve center of the new system of leadership.

science made possible a crude form of Marxism which dismissed all that fell outside this trinity to an unreal veil to be removed by revolution. With Taylorism, socialism could be reduced to positivism.

We are not saying that scientific management caused all this. The dream of encompassing technology and work within science was foreshadowed in Enlightenment philosophy and was hinted at by Francis Bacon. The trinitarian vision of synthesis would have blossomed in the twentieth century even if Frederick Taylor had never existed. But Taylor was the first man to convince the world that such a synthesis had in fact been achieved. Most of Taylor's concrete recommendations were rejected by his successors; however, his basic insistence that there existed one best way of managing industrial production, and that this way could be identified by applying the methods of natural science to the field of management, remained common assumptions among both friends and foes of twentieth-century modernization.

Taylorism in National Contexts

The American engineers' struggle for professional autonomy can be said to have had two main elements. One was the early-twentieth-century alliance with science, which aimed at strengthening the historical role of the engineer. The other was scientific management, which attempted to combine the historical and the managerial role within one set of knowledge practices. Taylorism implied reorganizing corporate bureaucracy in a way that enhanced professional autonomy. An organization was described through the metaphor of the machine, thus legitimizing the managerial role and offering engineering as an alternative to the nascent social sciences.

During the early 1930s Thorstein Veblen's book *The Engineers and the Price System* became the bible of the technocratic movement. (See Jamison, chapter 4 in this volume.) Veblen was inspired to write the book by observing the political activities of H. L. Gantt, a close associate of Taylor. Veblen's book was technocratic in a double sense. Veblen tended to look at production from the point of view of the *oikos*, reducing the capitalistic dimension of modernity to the industrial and overlooking the social conditions of production. In Marxist terms, one could say that Veblen did not speak for replacing capitalist relations of production with new relations of production; rather, he saw the relations of production as being somehow unreal and wanted to reduce them to the forces of production. Veblen was an outsider to the American society, and his was a somewhat "un-American" form of technocracy.

Americanism as a discourse was technocratic in a different sense. It called for the scientific professional's anonymous rule through the business corporation. This vision of the professional "technate" was also present in Veblen, and in a mild sense it inspired American reformers from progressivism to the New Deal. When it emerged on the European intellectual scene, however, it inspired technocratic visions that were different from the American ones.

In England and Germany, Americanism threatened the hegemony of the ruling classes by devaluing their cultural capital, but in France Americanism strengthened the dominance of the modernizing elites in Paris over Catholic and provincial France. The Eiffel Tower, built for a world exhibition in 1899, signified the successful union of high technology and high culture. The reason Americanism did not pose a threat to the French ruling classes is that, in the nineteenth century, technology had been appropriated within a cultural discourse that was itself naturalistic—namely, the doctrine of classical positivism, the intellectual creed of the French republicans. Americanism in general and scientific management in particular seemed to fit this discourse well. There were, however, some important differences, which led to a process of reinterpretation.

Comprehending French classical positivism is more or less synonymous with comprehending the peculiar position of the engineer in the French national mentality. The French and the German engineering professions were spiritual children of the French Enlightenment. There was one important difference between the French Enlightenment and the two other great Enlightenment cultures, the Scottish and the German (Macintyre 1984). In Scotland and Germany, radical Protestantism—Presbyterianism in Scotland, pietism in Northern Germany—opened the universities to the Enlightenment. In Scotland and Germany, the Enlightenment was articulated by university professors of philosophy, history, and theology: Ferguson, Hutcheson, and Adam Smith; Kant, Wolff, and Friedrich Schiller.

In Catholic France, church control kept Enlightenment culture separate from the academic humanities. Enlightenment culture became the domain of natural scientists and free intellectuals, giving the French Enlightenment its radical and naturalistic bent. The Revolution shut down the deteriorating Catholic universities, but they were not replaced with new seats of learning until 1806, when Napoleon created his "university." The Napoleonic "university," which remained in effect until the end of the nineteenth century, was, in the words of Fritz Ringer (1982: 111), "not an advanced institution of learning, but an hierarchically organized and centrally controlled corporation that encompassed all public secondary and higher education and its personnel." "Napoleon's conception of education," Ringer notes, "was narrowly practical."

The impact of the Revolution on academic literary culture was largely negative. French intellectual life, even French science, tended in the nineteenth century to flourish everywhere but within the academic institutions of the state. Regarding the scientific culture the Revolution did in fact have a positive program. The Jacobins wanted to mobilize natural science in the service of the nation. In the year of the Terror, 1793–94, they established the first engineering university in the world: the Ecole Polytechnique in Paris. The Ecole Polytechnique was a small institution with a great impact. According to Ringer (1982: 70ff.), an astonishing 9 percent of all persons mentioned in French contemporary biographical dictionaries between 1830 and 1960 were graduates of this institution, which produced about 100 engineers per year.

The Ecole Polytechnique differed from later, similar institutions in other Western countries in that, to a large extent, it prepared its graduates for the higher positions of the civil service. Until recently, Germany and the Scandinavian countries were ruled by jurists, while the English civil service was staffed with Oxbridge gentlemen educated in the liberal arts. In the French civil service, however, engineers were perhaps the most influential professionals. The Ecole Polytechnique set a model both for French engineering education and for the French concept of science. It stood for a scientistic concept of engineering science, emphasizing natural science, the main didactic principle being mathematical drill. This scientism was copied by later high-level engineering institutions such as the Ecole Centrale des Arts et Manufacture (founded in 1829) and thereby brought into industrial practice. On the other hand, in striking contrast to twentieth-century English engineers, French engineers of the nineteenth century prided themselves on being engineers rather than "pure scientists." The unique thing about nineteenth-century French engineering is that the knowledge practices of the academic engineers were hegemonic in relation to those of the natural scientists. In the Anglo-Saxon countries, engineering was to drift toward natural science. In France, natural science drifted toward engineering.

Thus the Ecole Polytechnique produced and reproduced an understanding of science that was rationalistic in its theoretical emphasis but at the same time strictly utilitarian. Science was in the service of industry and the state. Above all, since both of the main centers of French nineteenth-century scientific learning—the Ecole Polytechnique and the science department of the Ecole Normal Superieure (the "teachers college")—had as their main function to educate leaders of the civil service, science in France took on a distinctly political dimension. Science didn't serve primarily "autonomous learning" or private industry; rather, it was the knowledge practice of those who ruled the state.

The two national characteristics of French engineering education—that it prepared people for jobs in the civil service and that it was on a par with the natural sciences in intellectual prestige—led to a third national peculiarity: French engineering tended to understand itself more as a kind of social science than as a kind of natural science. The Ecole Polytechnique mode of knowledge production formed the discursive framework out of which grew French nineteenth-century positivism, and, with it, the dream of the non-political state, ruled by formalized but utilitarian science.

Henri de Saint-Simon, an eccentric ex-count and a hero of the American war of independence, the man who first used the term "positivism" and the first to use "avant-garde" and "industrialism" in something like their modern senses, set up his residence near the Ecole Polytechnique in 1798 in order to be better able to study the meaning of techno-scientific progress in relation to the march of history. It was at the Polytechnique that Saint-Simon met the young engineering student August Comte, who became his pupil and secretary. Comte later in life found his only long-lasting academic engagement at the Polytechnique, and Comtian or Saint-Simonian positivism remained the pervasive intellectual current at the school through most of the nineteenth century.

The continuing conflict with the Catholic intellectuals meant that there emerged in France a relatively clear dividing line between the moderns and the anti-moderns. When literary culture differentiated itself from the economic bourgeoisie, it often did so by being more modern than the bourgeoisie—the artists and writers through the cult of the "avant-garde," the university professors by being fiercely republican and anti-clerical. France was unique in developing a homogeneous bourgeois culture, common to economic, political, and cultural capital. Having received a classical education through the baccalaureate, the high-status academic engineer was able to act as a mediator and link between money and culture, between industry and the arts. In France, science, technology, capital, republican politics, and "enlightened culture" united in a radical project of modernity.

French-Style Americanism

Americanism came to France at a crucial period in time, just as the French intellectual field was about to change to a more Germanic ideal type. If there had been a turning point in the French modernization process, then it was 1870–71, the winter of defeat. After the disastrous German war, the republicans set out to restore the international prestige of French learning, which had suffered badly in the era of the Napoleonic "university." A series of reforms, culminating in 1896, established German-style universities, with philosophy and the academic humanities situated near the top of the status hierarchy. To the French, Germanization represented modernization and provided a means of "catching up with" the leading powers.

When Henri Bergson was made a professor at the Sorbonne, in 1900, it was a sign that some of the strong points of the humanist intellectual tradition were going to be reintegrated into the mainstream of French academic life. However, at that time Americanism met little audible opposition from the emerging class of humanistic mandarins. The doctrines of Frederick Taylor and Henry Ford were instant successes among republicans (receptive as always to the new and modern) and was endorsed by politicians like Albert Thomas and Gustave Herriot, by artists and intellectuals like Fernand Leger and Le Corbusier, and by engineers and businessmen like André Citroen, Louis Renault, Le Chatelier, and the Michelin brothers. They typically underscored the French intellectual roots of Taylor's method:

The human spirit must submit to a scientific method. . . . This method is, as you know, neither American nor new, she is French originating in our seventeenth century, it was Descartes, inspired by Socrates, who invented it, and it accords so well with the needs of human reason that it has triumphed everywhere. (André Citroen, quoted in Schweitzer 1995: 20–21)

The practical impact of Americanism in French industry during the interwar years was less significant than the ideological, even though it was the French who were the European pioneers in the application of both scientific management (Michelin) and the Fordist assembly line (Citroen) (Fridenson 1978). Americanism was endorsed, but it was reinterpreted to fit existing French management theory, such as the Fayol system. In the Anglo-Saxon context scientific management is often identified with bureaucratization. In the French context, such an identification would be misleading. The Taylor system was in certain ways anti-bureaucratic. It was formulated in opposition to the linear and hierarchical "army model" of organization, common in industry before Taylor. Taylor's scientific management was bureaucratic in the sense that he insisted that all work be planned in advance and written out by the planning department of the engineers. Taylor's organization charts were, however egalitarian, rather than hierarchical. He believed in "functional fore-manship," according to which each "function" in the production pro-cess was to be supervised by a foreman who reported directly to the organization's top layer, which was also functionally organized on an egalitarian model. These "democratic" aspects of scientific management did not fare well with the French engineers. In 1925 a celebrated reconcil-iation meeting took place between the French Fayolians and the French Taylorites. On the face of it the score was settled in favor of the Taylorites. Fayol declared that his and Taylor's systems "were the same" and that he was a Taylorite. What actually happened in French scientific manage-ment was that Taylorite functionalism was relegated to the lower levels of the organizational charts, while French-style hierarchy prevailed in the upper echelons of the factory (Merkle 1980: 165).

Also, in the general use being made of Americanism in providing technocratic solutions to the capital-labor conflict, the French engineers tended to remold the promises of earlier French models. Among the engineers of the French civil service, Taylorism and Fordism regenerated the old, half-forgotten Saint-Simonian dream of a United States of Europe, ruled by an engineer bureaucracy working within the framework of a science of administration (ibid.: 136–137). In the interwar years, and under the Vichy government, technocracy was a major factor in French political and cultural life. However, in France technocracy did not primarily mean the invisible rule of the corporate professionals envisaged by American thinkers. The technocracy feared or hoped for among French intellectuals was, rather, the visible rule of the powerful state technocrats (Zeldin 1981: 276–318).

As the new class of humanistic mandarins rose to power in the late 1920s and the early 1930s, anti-Americanism became fairly common among French intellectuals.[4] Anti-Americanism was almost exclusively a property of the right, often the extreme right (Lucier Romier, Alfred

4. "Writers who visited the United States in the early 1920s had looked rather favorably upon what they considered at the time a symbol of efficiency and progress. But within ten years the assembly line had become the symbol of something else: American materialism, mass society, indeed a novel form of collectivism that the most radical commentators, such as Fabre-Luce and Romier, likened to Bolshevism. Their slogan was 'Neither Ford nor Lenin.' " (Boltanski 1987: 107–108)

Fabre-Luce, Paul Morand, Henri Massis). On the left, "modernists" such as the liberals Hyancantie Dubreuil and Emile Schreiber and the socialists George Boris, Robert Marjorlin, and Pierre Mendès-France took a passionate interest in the New Deal. In 1930s France the opposition between modern and antimodern was still analogous with the opposition between left and right.

In practice, French society remained relatively mildly touched by Americanism in the interwar years, despite the widespread enthusiasm for Taylor and Ford. In the social sciences, another important aspect of Americanism, the French had made a head start with Durkheimianism round the turn of the century. In the interwar years, Germanization of the academy meant stagnation for the Durkheimians. The social sciences were only fully institutionalized in the 1960s. Only then did the Ecole Polytechnique lose its status as the most meritorious school for candidates for the higher spots in business and the civil service, being finally overtaken by the social science Ecole National d'Administration. Here at last France was being Americanized.

The standard criticism of classical as opposed to logical positivism is that its ill-conceived dream of the scientized polity stemmed from its failure to differentiate properly between "the is" and "the ought." Differentiating science from politics did not come naturally in France, where science was technology and technology was educational preparation for government. Also, there were argumentative reasons why the French classical positivists didn't make the distinction—reasons having to do with the premises of rationalism.[5] British empiricism, unlike French rationalism, thought mind and reality only coincidentally related to

5. It was in the 1830s and the 1840s, at the heyday of classical positivism, that Descartes was rediscovered and established as the founding figure of modern French philosophy. Descartes had not been a major figure in the French Enlightenment. The *philosophes* had looked to England, to the empiricism of Locke and Newton, for inspiration. Two intellectual currents were responsible for the canonization of Descartes in France. One was Comte and positivism; the other was Cousin and eclecticism. In 1837 Victor Cousin, Louis Philippe's Minister of Education, published what would long serve as the textbook for the newly introduced baccalaureate course in philosophy. Ironically, Cousin had taken the idea of making Descartes the founder of modern philosophy from Hegel. In his history of philosophy, Hegel had endowed Descartes with an overwhelming narrative function, letting Descartes father modern thought along with the very dichotomies between subject and object, thought and being, that the Hegelian dialectics were destined to transcend (Rørvik 1993; Zeldin 1977: 114).

each other. From David Hume to George Moore, subjective values were opposed to empirical matters of fact. To Moore the is was reality, whereas the ought was a "non-natural property." The British tradition tended toward making the distinction between the is and the ought a distinction between the real and the non-real, the subjective and the objective. Within the premises of German idealism, on the other hand, the is would at best be an unfathomable *Ding an sich*, to be handled by means in the services of ends. Among German intellectuals, the distinction between the normative and the descriptive became a distinction between means and ends. It was Kant who insisted that this was an absolute distinction. He separated means from ends in order to make room for human freedom in a deterministic Newtonian universe.

Kant did not have the technology question in mind. It was, however, into the means-ends distinction that the neo-Kantians Max Weber and Werner Sombart appropriated technology when "the question of technology" arose in the years before World War I. In France the means-end dichotomy rang less true, the cultural conflict between technologists and humanists (the suppliers of means and ends, respectively) being less pronounced than in Germany. The positivist tradition saw technology as embodied enlightenment—not just as a means, but as an end in itself.

Twentieth-century logical positivism arrived at a third way to separate the is from the ought. It termed the distinction a logical one. Henry Poincaré argued that the ought occurs in imperative statements, the is in indicative statements, and that if the premises are in the indicative one surely cannot deduce an imperative conclusion. This kind of neo-positivist criticism was one reason why classical positivism was slowly losing its grip on the French intellectual field in the early decades of the twentieth century. Equally important was the growing influence of Bergson and German idealism. Besides, the evolutionist premises of classical positivism were undermined by the calamities of World War I.

Scientific management reinforced the weakening hegemony of classical positivism and extended its life span somewhat. One of Taylorism's sources of intellectual appeal was precisely that it did not separate the is from the ought. Taylor's reasoning proceeded in three steps. Step 1 was to determine the best way in which a job could be done. Step 2 was time study. Taylor thought he determined not only how quickly a job could be done but also how quickly the job should be done (Layton 1971: 141). Finally, Taylor meant to be able to measure the contribution of each man and each factor of production. From this the scientific manager could determine the just wage. Taylor thus meant to have eliminated the entire field for ethical and political controversy between labor and capital. Lifted from its origins in the "stateless" states of America, scientific management brought back the French dream of the scientized polity.

The Engineers in the German *Sonderweg*

After Carl von Ossietsky was awarded the Nobel Peace Prize, Adolf Hitler made it a legal offense for German citizens to accept Nobel Prizes. In compensation, Hitler in 1938 introduced a "national prize for art and science." The Nazi prize, however, was not given to scientists. Hitler's 1938 prize was awarded to four engineers, men from practical life, rather than the abstract realm of science: Willy Messerschmidt, Ernst Heinkel, Fritz Todt, and Ferdinand Porsche. That same year Hitler also presented the American Henry Ford, on his 75th birthday, with Germany's highest decoration for "outstanding services to mass-production" (Collier and Horowitz 1987: 178). This was also the year that Hitler launched the Volkswagen project, copying the production concepts of Henry Ford (Siegfrid 1988). Few historians have noticed these events, but we believe they reveal something significant about the discourse on technology and science within the German *Sonderweg.*

In his 1969 book *The Decline of the German Mandarins,* Fritz Ringer gave a new twist to the long-standing debate on the *Sonderweg.*[6] Ringer argued that the German trajectory was constituted through the combination of an early educational reform and a late industrial modernization. In eighteenth- and nineteenth-century Protestant Germany, a relatively efficient, well-funded, and socially inclusive educational system endowed a humanistically educated elite with much social power and prestige. With late-nineteenth-century industrialization, bitter social rivalry developed between this bourgeoisie of culture and the rising bourgeoisie of business. German intellectuals variously worded that rivalry as a conflict between "culture" and "civilization," between "state" and "society,"

6. The *Sonderweg* debate was a prime concern of German scholarship up until World War II. The crimes and disasters of the Third Reich gave rise to a new, international debate on the German *Sonderweg.* Common to both debates was the concern over whether Germany had had its own mode of development— one different from "the countries of the West" (France, England, and the US). After the breakdown of modernization theory, a consensus seems to have developed that every country, not only Germany, had its *Sonderweg.*

or between *Gemeinschaft* and *Gesellschaft.* This "civil war" within the German middle classes put the engineering profession in a precarious position.

The German engineering profession was "scientized" much earlier and more thoroughly than either the English or American engineering profession. In the first half of the nineteenth century, engineering colleges (*Hochschüle*) were founded in several German states, on the French model. There were, however, important anti-scientistic elements within the discourse of the German technologists. First, because of the peculiar position of the academic humanities within the German discourse of science (*Wissenschaft*), the German concept of science was not scientistic. The summit of scientific prestige was to be found in the philosophical and historical sciences of interpretation. Whereas science, in the Anglo-Saxon world, had become a methodological concept meaning knowledge obtained by the methods of the natural sciences, *Wissenschaft* meant "learning" or "scholarship." Second, engineering was established in independent institutions for engineering education, outside of the "normal" universities. This facilitated the cognitive independence of engineering in regard to the natural sciences. Engineering was recognized in Germany as a science by itself, rather than as "applied natural science." The institutional split between natural science and engineering was strengthened by the difference in social status between the natural scientist and the engineer. Unlike the engineer, the natural scientist was securely placed within the cultured elite.

Relative to the Anglo-Saxon universities, the German engineering universities were closely tied to the state. Still, the Humboldtian ideals of freedom of research were not merely rhetorical or ideological. The German institutions did have a sizable amount of autonomy in regard to the state, especially relative to the French institutions. What developed in German engineering education, instead of the Anglo-Saxon profession-education relation and the French state-education relation, was a net of close relations between industry and education. German academic engineering education was sensitive to the needs of industry, even though didactic practices were very theoretical compared to the conditions that still prevailed in nineteenth-century Anglo-Saxon engineering.

In short, the important fact is that German engineering at an early date was institutionalized according to a tradition of *techne* rather than one of scientism. Engineering was seen not as a subdivision of natural science but as an independent form of knowledge in its own right. The German engineering discourse identified with what we have termed the historical role of the engineers as toolmakers rather than as managers. The engineers were chief recruits to management positions in German industry. German firms, however, were less bureaucratized than their French counterparts. The Germans spoke not of "management" but of "leadership" (*Leitung*), emphasizing functional knowledge and charismatic qualities. In France the manager was a "patron"; in Germany he was a *Führer* (a first among equals), supposedly distinguishing himself through technical competence and inborn leadership qualities.

The German engineer lacked the academic prestige of the French engineer. He did not have access to the higher echelons of the civil service, which remained the domain of the jurists. The French Polytechnique engineers were bearers of a hegemonic discourse in relationship to both natural science and the humanities. The French engineer was a major definer of "the project of modernity." The German engineer lacked this power and had to legitimate himself by other strategies (for instance, by proving his usefulness to industry).

In France the academic engineer usually lived off the state, while the intellectual lived off the market. In Germany it was the other way around. University professors, particularly within the academic humanities, were a semi-aristocracy in nineteenth-century Germany. According to Ringer (1982), 45 percent of all persons mentioned in German biographical dictionaries between 1830 and 1960 were university professors. The comparative figure for France is 16 percent. On the other hand, "free intellectuals" constituted 24 percent of the French elite, versus only 13 percent of the German elite (Ringer 1982: 70–71). The German engineer attempted to move in an intellectual field dominated by the mighty mandarins, the academic humanists. The social hegemony of the mandarins made it advantageous for German engineers to try and gain cultural acceptance by presenting technology as a cultural construct within culture rather than outside it. (See Hård, chapter 3 above.) Unfortunately, the German discourse of cultural technology was tied in with the ever-more-aggressive nationalism of Wilhelmian, Weimar, and eventually Nazi Germany.

In Germany there was some initial enthusiasm for scientific management before and during World War I (Merkle 1980; von Freyberg 1989). The engineer Wickard von Moellendorf, a close associate of Rathenau, was an eager Taylorite. Inspired by his own and Rathenau's successful guidance of the German war economy, Moellendorf lifted his planning horizon from the firm to society, introducing the public to the notion of a *Planwirtschaft*. Generally speaking , however, scientific management was dismissed as a management theory. It was at odds with the underlying grammar of German engineering discourse: *techne,* the historical role, technology as culture. Also, scientific management was unacceptable to the anti-positivist discourse of the mandarins. It was dismissed as yet another example of a rationalistic and alienating civilization. But fascination for the technological aspects of Americanism was strong. As a result, scientific management was to a certain extent imported as technology rather than as management theory. The much-publicized *Rationalizierungsbewegung* of the 1920s, became, however, mostly a movement for further concentration of the already heavily cartelized German industry.

In the early years of the Weimar republic there developed in Germany a politically influential technocratic discourse of *Gemeinwirtschaft* in the writings of the social democrats Karl Kautsky and Rudolf Wissel, the liberals Walther Rathenau and Rickard von Moellendorf, and the Nazi ideologue Gottfried Feder. These technocratic visions of the Weimar Republic differed from their American and French counterparts in several ways. First, they focused very specifically on the engineer. One reason for the German focus on the engineer was the strength of organized labor. In a society on the brink of revolution, the strategic placing of the engineer in industry—between labor and capital—made him a focus of interest to reformists and reactionaries alike. Unlike in France and the United States, the German technocratic visions had certain anti-scientistic traits, more pronounced the further to the right on the political scale one moved. The engineer was not, as in the United States at the time, just one in a whole family of "applied scientists." Second, and an aspect that distinguished German from French but not from American technocracy, was its anti-statist nature. Neither Moellendorf, nor Rathenau, nor even the Marxist Kautsky believed in expanding the involvement of the state bureaucracy in the economy (Rathenau 1917a: 28ff.; Rathenau 1919a: 87; von Moellendorf 1919: 1ff.; Kautsky 1919: 6). At stake here was a conflict of jurisdiction between jurists and engineers. State bureaucracy in the German trajectory meant juridical bureaucracy of the Weberian type. For the power of the engineer to be increased, the activities of the state had to be curtailed and the autonomy of industrial production had to be safeguarded. *Gemeinwirtschaft* was a vision of transcending the liberal economy and the antagonism between the classes by corporatism: vertical integration between firms, horizontal integration of workers, engineers and owners in common unions. Finally and most importantly, German-style technocracy was distinguished by its anti-capitalist nature. In French and American technocracy the scientific professional was seen as a potential mediator between "industrialism" and "capitalism," set to correct the misdoings of capitalism but not to abolish it. German-style technocracy, even in the non-socialist versions of the liberals Rathenau and Moellendorf and in the socialist version of the Nazi Feder, was skeptical of the modern, differentiated economy. The mutual interdependence of technological society and the organic solidarity of factory work would make possible the return to a concrete and unitary mode of living. Affluence would eliminate class differences. Technology would break the alienation of Mammon and lead man back to his roots. Ernst Haeckel and others even launched a science of the *oikos* (ecology) promising technological man's reunion with Mother Nature.

The persuasiveness of the technocratic vision stemmed from interaction between the grammars of German technological and humanistic discourse. Anti-capitalism was made credible by the German emphasis on the historical role of the engineer as tool producer rather than as manager. The humanists supplied the language of the *Gemeinwirtschaft*, drawing on Greek sources. Aristotle's normative distinction between production for use and production for exchange loomed large in the discourse, as did his concept of *autarkeia* (self-sufficiency, later to become the key term of Nazi economic policy).[7]

For all its explicit anti-positivism and anti-scientism, German technocracy attempted to make the cultural symbol of the engineer do what it had previously done in France: reconcile cultural and economic capital— "culture" and "civilization." In Germany, where this division within the middle classes ran deeper, "culture" was stronger and technocracy became a "reactionary modernism" (Herf 1984).

The German engineers were politically divided in the Weimar years. Frustrated by the political passivity of the Verein Deutscher Ingenieure (the main German engineering association), a section of engineers in 1918 left and formed the Reichsbund Deutsche Teknik (RDT). The RDT campaigned widely but unsuccessfully for the establishment of a German ministry of technique that could give the engineers a say in government (Ludwig 1974).

The most political group within the RDT was led by Gottried Feder.

7. On production for use versus production for exchange see *Politics* I ix and x. In Aristotle *autarkes* is the *telos* of the state. "The final association, formed of several villages, is the state. For all practical purposes the process is now complete; self-sufficiency (autarkeia) has been reached." (*Politics*, I ii, 1252b27)

Like the rest of the RDT, Feder pressed for the economy to be reformed in keeping with the imperatives of the "engineering mentality"; however, Feder dressed his propaganda in a violently anti-capitalist and anti-Semitic language. Feder's group of technocratic engineers was, in the 1920s, a key element in the Nazi Party. A statistic compiled by Jarausch (1990) shows that half of all the engineers who sat in the Reichstag between 1918 and 1933 sat for the Nationalsozialistische Deutsche Arbeiterpartei, and that most of them came from RDT and/or the Feder group.

In a key passage in *Mein Kampf*, Hitler (1925: 199) describes an economics lecture for demobilized soldiers he had attended in 1919:

Previously I had not seen clearly enough that pure capital calculated as a result of productive work is different from the capital which is derived from pure speculation. . . . Now I was told that from one of the gentlemen who lectured at the course I mentioned: Gottfried Feder. For the first time I heard a fundamental exposition on international finance capital. Having heard Feder's lecture it struck me like lightning that I discovered one of the most important platforms to found the new party on.

The German engineers were, generally speaking, not more susceptible to Nazism than those in other middle-class occupations. But in the 1920s Gottfried Feder, with his peculiar version of "the engineering mentality," was the most influential economic theoretician of the Nazi Party. After associating with the "left wing "of the NSDAP, he lost all influence after 1934. Feder was not an admirer of Taylor. He did, however, study with interest the writings of two other Americans: Thorstein Veblen and Henry Ford (Feder 1934: 84–85; cf. Hughes 1981).

Generally speaking, Europeans tended not to see much difference between the doctrines of Taylor and those of Ford. Ford's assembly line and five-dollar day were lumped together with scientific management, standardization, interchangeable parts, the giant corporation, and so on under the heading of Americanism. In Germany, however, the difference between Ford and Taylor came to loom large. In 1924 Friedrich von Gottl-Ottilienfield, an economist and a leading philosopher of technology who later joined the Nazi Party, wrote a pamphlet comparing Ford and Taylor. Ottilienfeld regretted that people often spoke of Ford and Taylor as if Taylorism and Fordism were the same, when they were in fact as different as "fire and water." He wrote that Taylorism ended in an "orgy of organization," destroying the soul of the worker and turning the factory into "soulless clockwork." Fordism, on the other hand, saved the dignity of work by leaving room for individuality within an organic whole (von Gottl-Ottilienfeld 1924: 8-11). *Taylorismus* became a negative term in the 1920s, while *Fordismus* generally had positive connotations. This seems odd in view of the picture of the German intellectual field as being dominated by romantic, "anti-mechanistic" notions. What could be more mechanistic than the Fordist assembly line?

The paradox resolves somewhat if we leave behind the abstract notion of Fordism and look into Ford's writings. In Ford's best-selling (ghost-written) books we find a curious blend of conservatism and populism. Paternalism merges with the call for charismatic leadership, efficiency with distrust of finance capital, *laissez faire* with anti-Semitism.

Ford and Hitler

Henry Ford claimed to speak for Americanism, which he proclaimed to be the goal toward which all civilization was striving (Ford 1926: 258). Yet Ford, as he portrayed himself, was in certain ways outside the mainstream of American thought on technology and society. He was a self-taught man, without formal schooling. His writings on the development of technology were far removed from scientism. Other American corporations in the early decades of the twentieth century tried to modify antitrust sentiments by eagerly emphasizing the "scientific" nature of their activities. Ford, on the other hand, boasted: "We do nothing at all in what is somewhat ambitiously called research, excepting as it relates to . . . our own particular function which, to repeat, is making motors and putting them on wheels." (ibid.: 55)

Ford depicted a world in which the machine was king. But he distrusted the scientific professional. On Ford's assembly line the movements of the worker were dictated by the machine, not by the scientific manager. In the Taylor system every man was to have his meticulously specified function in the work process, spelled out every day in written instructions. Ford detested written instructions and all other "red tape," trusting instead in "natural selection." Tasks were distributed to groups of two or three men, who were to decide on internal rank according to the principle of the survival of the fittest. In a curious mix of democracy and old-fashioned paternalism, every man, regardless of rank, was in principle allowed to bring his suggestions and complaints to the ears of the omnipotent director, Mr. Ford himself. The "Sociological Department" of the Ford factory was, on the other hand, little more than a police organization, entrusted with surveying the drinking and the sex lives of the employees.

Ford's economic philosophy centered on the distinction between the industrialist and the capitalist. In an almost moving passage, the richest man in the world described the plight of the industrialist in the days when the labor movement created a stir (Ford 1926: 28). In those days, wrote Ford, an employer was said to be the same as a capitalist. The fact, however, was that the industrialist was in the hands of the capitalist, industry being run on borrowed capital. The industrialist, Ford complained, was hardly able to get anything done, squeezed as he was between hostile workers and capitalists who wanted to plunder him. On top of it all, he had to accept being abused as a "capitalist."

Ford had long since determined who the wicked capitalists were. The keepers of unproductive capital who sabotaged the workings of the productive industrialists were the Jews on Wall Street. Ford's attempt at making a political career in the United States was a failure, but his book *The International Jew*, first published in 1920, was a success—in Germany. The German edition of *The International Jew* went through 29 printings before 1933.

The keystone of Ford's business strategy was autonomy. This meant, first of all, full independence from "finance capital." Ford systematically reinvested his profits so as not to have to depend on banks. Second, it meant control over the entire chain of production. Ford owned his own steel mills, iron mines, and rubber farms. Even the workers' food was supplied from the company farms. Ford's River Rouge factory—100,000 employees under the command of one man—was close to being a self-sufficient city-state.

This autarkeia, in the end, proved to be an illusion. When sales dwindled as depression struck in 1930, Ford countered by increasing production. He attempted to conduct "Keynesian" demand inducement at the level of the firm. The operation was a failure, driving the company close to bankruptcy. Ford had to give up his legendary high-wage policy, and, with his private army fighting ever more violently to avoid unionization, his factories came to resemble concentration camps. Disillusioned with the machine, Ford started writing on how to return the factory to the soil.

Herf (1984) has drawn attention to the impact of "reactionary modernism" among German academics in the interwar years. He argues that experiences of World War I as a technological war facilitated a blending of technological modernism with the romantic, nationalistic, and antimaterialist themes of the mandarin tradition. Reactionary modernism found support among the academic humanists and free intellectuals and among engineering professors. Herf claims that this paradoxical synthesis of the modern and the anti-modern and of civilization and culture created by the new right of the Weimar years provides a key to understanding the Nazis' attitudes toward modernization.

Frederick Winslow Taylor was too closely aligned with "the values of civilization" to be made into a persuasive symbol of cultural change in Germany. It does seem odd that the thoughts of a semi-illiterate mechanic from Wayne County, Michigan, should be better suited for symbolizing the reunion of culture with civilization. Yet, within the limits set by ignorance, Henry Ford did indeed portray himself as a true reactionary modernist. The man who, when questioned on what was the year of the American declaration of independence, famously proclaimed that "history is bunk" was a passionate protector of America's cultural heritage, spending vast sums of money on restoration and on museums to preserve the America "as of Mr. Edison's birth" (Ford 1926: 229). At the Ford Motor Company's research institute in Dearborn, Ford furnished a ballroom for 70 couples to which the research engineers were summoned twice a week for dancing practice. "Everyone," said Ford (1926: 227), "has to learn to dance in absolutely the correct way." In the 1920s the Ford Motor Company had 50 aging fiddlers on its payroll. Ford cooperated with General Electric in preserving American folk music, mobilizing for this purpose the most advanced recording technology available.

No one who has read Henry Ford's books of the 1920s can help being struck by the similarities to *Mein Kampf*: the unlettered style; the puzzling mix of ignorance and knowledge; the rampant anti-Semitism coupled with a decrying of "unproductive finance capital"; the search for a middle way between capital and the worker; the love for discipline and organization coupled with a distrust of formal titles, red tape, and professional expertise; voluntarism coupled with a puritanical craze for "self-sufficiency"; and, perhaps especially, the admiration for technological progress coupled with a nostalgic romanticizing of the past.

The assembly line and the conveyor belt have become arch-symbols of mechanized and alienated modernity. Still, because it lacked Taylorism's scientistic aspects, the Ford idea could be reinterpreted as a *Gemeinschaft* vision. German ideologists and intellectuals of the right and the left did just this in the 1920s. Ford's *Gemeinschaft*, his *oikos*, was the factory—thousands working side by side united under one leader. Hitler's *Gemein-*

schaft was the nation, but substitute the nation for the firm and Ford's vision for tomorrow becomes very much like the one spelled out by Hitler. These parallels were not primarily due to Ford's and Hitler's having studied each other's ideas, even though Hitler did read *The International Jew* before writing *Mein Kampf*.[8] The visions of Hitler and Ford sprang instead from similarities of "habitus" and of personality. Both men were self-made and untutored. They subjected themselves to repressive personal regimes: they were vegetarians, and, like Taylor, non-smokers and non-drinkers. From their backgrounds they carried with them a distrust, not only of socialism and capitalism, but also of science and education, and yet each developed a "progressive" world view organized around admiration for modern technology.

Final Remarks

The NSDAP cast its rhetoric within the premises set by the German discursive framework. Key elements of the mandarin traditions—the distinction between science and technological science, the rhetoric of *Gemeinwirtschaft*—can be found in Nazi discourse. Nazi ideology spoke the language of "culture." Yet Nazi rule was "modernizing" in letting new "uncultured" men into power, breaking down the established status hierarchies. Great mandarins such as Martin Heidegger and Carl Schmitt typically supported the Nazi regime in its early years, then turned away from it as the "uncultured" nature of Nazi rule became apparent.

Ringer's *Decline of the German Mandarins* has been interpreted as an argument for Anglo-Saxon science against German *Wissenschaft*. Ringer may be read as a belated representative of Americanism, arguing that the traditional institutions and ingrained values of the German academic community are politically suspect, having been partly to blame for the rise of National Socialism. We make no such normative claim. We believe, however, that historical research on the relations between Nazism and modernity would gain in concreteness and verifiability by focusing on Nazism's discursive relations with Americanism rather than with modernity in general. Americanism was, after all, the modernization project of the interwar years.

8. A *New York Times* reporter who visited Hitler in 1922 found "a book by Henry Ford" on his desk. A big photograph of the American industrialist hung on the wall. In the early editions of *Mein Kampf* Hitler praised Ford as the one great American entrepreneur who had not yet succumbed to the power of the Jews.

What is commonly known as Americanism was a controversial project within the United States. Americanism was an intra-American reform program that developed during the "progressive era" in response to American social problems and educational trajectories. Three aspects of Americanism loomed particularly large in interwar debates: "applied science," Taylorism, and Fordism.[9] In England, the power of Oxbridge literary culture and the strength of organized labor combined to limit the importance of the latter two aspects of Americanism. To the Germans, Fordism was the most amiable trait of Americanism. The French appropriated Taylorism, which, on the face of it, appeared to fit in well with the still influential premises of classical positivism.

In the early decades of the twentieth century the question of technology gained a political urgency it had never had before. The threat or promise of technocracy debated in the United States was the "invisible" rule of the impersonal scientific professions, the "technostructure" of big business. The technocratic vision debated in France was tied in with the statist and centralized traditions of French political culture. The link between science and political power was made more openly in France than elsewhere. Germans debated the relevance of Americanism to *Gemeinwirtschaft*, an anti-capitalist vision that centered specifically on the engineer. The predominance of this vision was made possible by the running conflict between economic and cultural capital in Germany and by the fact that the exchange value of economic capital was so low relative to that of cultural capital. Businessmen and technologists had to legitimate themselves in cultural terms. Modern technology and economy tended to be intellectually appropriated into some sort of regressive fantasy of returning to the imagined unity and stability of the pre-modern world.

9. A fourth component, the social sciences, did not gain public recognition as a key to Americanism until the postwar years. Often institutionalized as part of the Marshall Plan and other US-funded reconstruction programs, the social sciences became highly influential in most of Europe by the 1960s. For some interesting comments on English skepticism toward the social sciences in light of the late 1950s debate on "the two cultures," see Green 1974.

6

Swedish Grandeur: Contending Reformulations of the Great-Power Project

Aant Elzinga, Andrew Jamison, and Conny Mithander

Sweden as a Technological Power

The early twentieth century was a time when the various roads to modernity were staked out by intellectuals. Their task, among other things, was to interpret the new technological universe of electric utilities and science-based industry for the general public, to fit the new situation into recognizable frameworks of understanding. But they were also the self-appointed evaluators of the new symbols of progress, offering suggestions for controlling, managing, redirecting, even redesigning the new techniques in ways that could be socially acceptable. How were the new forces of mechanization and rationalization to be harnessed without causing widespread social disorder and anomie? In their attempts to formulate meaningful answers to this question, intellectuals drew on various cultural resources and were affected by various contextual tensions; but there were distinctive national patterns or discursive frameworks that helped to set the terms of debate.

In this chapter we shall examine the different meanings attributed to modern technology in the Swedish context, the symbolic significances assigned to it by various intellectuals. The main road of cultural appropriation in Sweden led to an interventionist, corporate state and a highly successful techno-economic regime—Sweden's famous "middle way" between capitalism and socialism (Childs 1936). But the so-called Swedish model, established in the 1930s, did not spring from nowhere; it represented a synthesis, or a constructive combination, of ideas about Sweden's reassuming its place as a great power that were first promulgated at the turn of the century as a nostalgic reaction to the century-long decline in the nation's fortunes.

It will be our contention that in the course of the interwar years an initially conservative framework of ideas was transfigured into a more comprehensive and widely acceptable social and political program that came to be rejected in its heyday by the followers of those who had first proposed it: the increasingly reactionary right of the late 1930s. The dissolution of the Swedish welfare state in recent years is, in turn, largely the work of the latter-day descendants of those "reactionary modernists" (Herf 1984) who now, in the 1990s, see the nationalism and great-power ambitions of the Swedish model as barriers to the inevitable globalization of industrial development and the integration of the Swedish political economy into Europe. As nationalism once moved from the right to the left, internationalism has since moved from the left to the right. But it is now the internationalism of a cosmopolitan capitalism; in the 1930s it was an internationalism of social democracy and labor solidarity.

The image of America was central to the contending visions of the great-power project that we will be discussing. Many Swedes had emigrated to the New World during the last third of the nineteenth century, and a multitude of popular images and beliefs about America had flourished in Sweden. Study visits to North America had also formed a part of the official investigation of the emigration problem. For conservatives at the turn of the century, the American work ethic and industrial prowess were the most important things to be imported to Sweden from the United States. For the conservation-minded, the United States provided organizational examples (national parks, the Sierra Club) as well as a certain pastoral ideal of industrial society—what Leo Marx (1964) has termed "the machine in the garden"—which was projected onto the vision of the future technological society; through them, a Swedish version of pastoralism, which is perhaps best thought of as a rural consciousness, worked its way into the social democratic synthesis. Leading engineering spokesmen and industrialists saw American corporate research, organizational efficiency, and scientific management as central elements in action programs for Swedish industry (Runeby 1978). For their part, the social democrats—who rose to prominence in the 1920s and then came to power in 1932 to begin a governmental reign of more than 40 years—appropriated these elements and gave them new connotations, marked, in part, by another set of influences from the United States, namely pragmatism and progressivism. In the 1930s, while the social democrats focused largely on the positive aspects of the United States, anti-Americanism was increasingly played up on the far right, spurred by an aristocratic contempt for the "vulgarity" of American culture and by political opposition to the peculiarly American combination of equality and individualism.

Historians and literary scholars like to refer to the Stockholm Exhibition of 1930—visited by some 4 million people—as symbolizing the onset of a second, more popular phase of modernity in Sweden. (See, e.g., Holm 1965; Kylhammar 1990; Nolin 1993.) At that exhibition, functionalist artists, writers, and architects called on their fellow Swedes to "accept" modern technology. (*Acceptera*, a book published at the time of the exhibition by a group of intellectuals, was widely disseminated.) The Stockholm Exhibition displayed a bright, mobile, and progressive future. Rationality and efficiency were the catchwords. Natural resources were to be used efficiently, and life, both in the home and in society at large, was to be organized rationally. "Down with the beautiful" and "In with order" were among the slogans of the day (Wennström 1986). Students and writers started up a number of new journals, most of them short-lived, to promote a variety of avant-garde ideas in art, literature, and social analysis. These journals had telling names such as *Fönstret* [*Window*], *Fronten, Spektrum, Atheneum, Karavan*, and *Presens*. Only *BLM*, a literary magazine sponsored by the Bonnier publishing house, has survived.

Within a year, dark clouds appeared. The gunshots that killed five and wounded five more striking sawmill workers in the Ådalen district of northern Sweden in 1931 provided a shocking reminder that class strife had not been eliminated by modernist utopianism. The following year, the suicide of the industrialist Ivar Krueger brought home the continuing significance in modern Sweden of the irrational and calculating—and highly unpredictable—nature of capitalism. Some of the literary intellectuals turned away from society to explore their own psyches; others longed for earlier times when life seemed to have been less complicated. The group around the journal *Karavan*, for example, reacted "against life's mechanistic intellectualization" and "wanted to work for more immediate, vital and instinctive forces of life." In a 1935 essay, the radical socialist Per Meurling cited the writer Artur Lundkvist more or less sympathetically: "'I am a peasant,' declared Artur Lundkvist with pomp. 'I believe in the archaic, the elementary. I believe in rocks and earth and rain and sun.' " (Meurling (1984: 30). Meurling ridiculed the "literary front" as "slightly loony individualists and aesthetic snobs." Their revolt, if it can be called that, was, however, not directed against technology, but rather against the constraints of the past. Lundkvist, who would later become one of Sweden's most influential literary intellectuals, saw his youthful revolt primarily in terms of modernism, and he often glorified the machine as an intrinsic part of the modern society:

. . . modernism was a passion for me, even more important than how my poems were judged. I wanted above all to be a representative, if one of the least significant, in a literary movement which I considered to be international and represent something new. . . . My glorification of the machine wanted to propose a solution to a general problem by portraying a future condition of harmony between man and machine, after both had found their proper place in the social totality. It seems to me to have been a more fruitful attitude than the usual, reactionary backward-looking machine criticism. (Lundkvist, quoted in Lindblom 1991: 36, 39)

Crister Skoglund, in his history of student radicalism (1991), points to two distinct but intermingling strands in the functionalism of the times: an aesthetic strand and a scientific one. The aesthetic strand contained a rejection of regimentation, while the scientific-rational strand fed on regimentation. The one side could therefore provoke and poke fun at mechanization, while the other side glorified facts and objectivity (Skoglund 1991: 189ff.). This tension was also articulated in political terms, as a tension between cultural radicalism and socialism. As the 1930s progressed, these two inclinations tended to pull intellectuals and politicians in different directions: toward individualism and cultural pessimism, as one pole, and toward collectivism and what Skoglund calls "conditional optimism" as the other. The eventual conditional utopia—Sweden's middle way—presupposed that "science and rational thought should lay the basis for decision making in a completely different way than before" (ibid.: 192).

The generally positive response to "the machine" at the time of the Stockholm Exhibition was inspired by a particular set of technologies, namely electricity, film, radio, the automobile, the airplane, and household appliances. These were seen by many intellectuals as providing the tools for a more standardized and popularly oriented form of technological civilization, based on brand-name products, functional design, and mass entertainment. These technologies thus appealed to egalitarian-minded intellectuals more than to those who professed one or another kind of elitist faith. The more exclusive conservative great-power project had, in its time, been a response to an earlier phase of industrialization: the phase of large-scale industry—characterized by Bessemer steel, machine tool production, steamships, and railways—which in Sweden exploded onto the national scene in an intensely concentrated way in the 1870s (Gårdlund 1942). It is important to distinguish the phases of industrialization from one another, since they rested on rather different technological regimes and since the ensuing intellectual debates thus involved the mobilization of somewhat different (or at least differently rearranged) sets of cultural resources.

What has been called the mythology of modern Sweden—the idea of Sweden as the first modern nation (Ruth 1984)—drew both on indigenous intellectual traditions and on internationally transmitted ideas. Much of its broad appeal derived from the ingenuity with which intellectuals managed to combine its multiple legacies, including great-power imperialism, social democratic internationalism, instrumental rationality, and a touch of rural populism. With pastoral symbols and a reliance on experts or "rationalizing intellectuals" (Eyerman 1985) to take charge of the process, modernity took on a distinctly Swedish style in which technological progress was to be accepted as the great equalizer, the main source of both economic growth and social welfare. As Löfgren (1993) has argued, the modernization of Sweden was facilitated by a new political rhetoric and by a set of social practices that fostered a "collective individual" sharing a common cultural repertoire, which included opinion surveys, modern health rituals, mass communications, and functionalist aesthetics in housing and industrial design.

In the seventeenth century, Sweden had controlled almost the entire Baltic region, from Bremen in the west to Dorpat in the east, as well as occupying most of the Scandinavian Peninsula. Sweden's great-power status had been founded on a combination of rich natural resources (mines and forests) and a militaristic and highly bureaucratic mode of governance. Even though the country's land mass diminished quite dramatically in the coming centuries, the great-power era left its stamp on the national political culture and, in particular, on the forms of interaction between intellectuals and the state (Eyerman 1992). It also left in the Swedish mentality a conception of national grandeur, a widespread belief throughout the society that the country was destined once again to play a leading role among nations.

In this chapter we will trace some of the trajectories through which this great-power project, in the first four decades of the twentieth century, was transformed from a primarily conservative and militaristic narrative into a more popular idiom, or ideology, of modernization. In its perhaps earliest and most explicit form, it was promoted by a self-proclaimed "young right" in Gothenburg, Sweden's second-largest city (located on the west coast). The role of the "young right" has been neglected in many historical accounts that have focused on the social democratic and liberal roots of the Swedish model of social engineering and technological rationality. Even though these conservative intellectuals were by no means as central to the modernity debate as their counterparts in Germany, there can be no denying that they played an important role in setting the terms of the Swedish national discourse. Whereas the "young right" drew on a rhetoric of national chauvinism and elitism, social democrats and liberals emphasized internationalism and equity. Though both sides of the debate were modernist, their points of departure thus lay in different value systems, and this was to have important implications for the symbolism that came to be ascribed to "the machine."

Reinventing Tradition: The Discursive Framework

More than any other event, the dissolution of the union with Norway in 1905 served to set in motion a Swedish "debate" about modern technology and its social implications. In reacting to Norway's independence (and to the long-term decline of their country's status in the world, epitomized for many by the massive emigration to North America), Swedish intellectuals mobilized a range of cultural resources to try to instill in their countrymen a new sense of grandeur or at least optimism. In the land of Linneaus, the technology debates had a decidedly pastoral air about them, and there was relatively strong interest in conservation and an efficient utilization of natural resources. Indeed, the vast northern forests were seen by many of the participants in the technological debate as providing the basis for an industrialization that could bring back to Sweden something of its lost prominence in the world. Norrland was to be the great frontier for expansion and development that the western territories had been for the United States (Sörlin 1988). In reality, Norrland never inspired much of an inner colonization, even though the forests did provide the raw materials for industrial progress. But geographical and natural-resource factors nonetheless contributed a rural stream of consciousness, especially on the conservative side of the political spectrum, as intellectuals sought to integrate machine technology into the natural landscape by giving it new kinds of meanings in both theory and practice.

Neither side in the debate ever pursued openly confrontational politics; both preferred to retain, and indeed strengthen, the paternalist consensus ideal that was a part of Sweden's political heritage. Above all else, the rationalist mode of negotiation and consensus building was characteristic of the Swedish style of debate, and it remains so even today. This brings several deeply ingrained features of Swedish history to mind.

First, Sweden's distinctively bureaucratic mode of governance derives from the imperial culture of the great-power era of the seventeenth century, when absolute monarchs created a multi-faceted state apparatus that unified the interests of aristocrats and peasants in a common project of expansion. In the words of Ruth (1984: 85): "The tradition of rational administration of the public interest is the paramount and never discarded legacy of the great-power era in Sweden." Also in the seventeenth century, the emphasis on order, discipline, and systematization that is so noticeable in the taxonomy of Linneaus, in the chemistry of Berzelius, and in the later innovations of Ericsson, Nobel, Dalén, and de Laval became a definitive component of Swedish science and engineering. (See Jamison 1982.)

A second feature is the *bruk*, the archetypal rural mining community out of which the economic development that laid the groundwork for Sweden's short-lived imperial expansion emerged in the medieval period. These enclaves of capitalism, which combined the extraction of silver, copper, and high-quality iron ore with metallurgical and munitions workshops, were overseen by aristocrats and served as conduits for the importation of foreign technical expertise (especially Walloons from the Low Countries). The *bruk*—which the German poet Hans Magnus Enzensberger has characterized as a micro-level precursor of the modern Swedish welfare state—was a paternalistic, semi-feudal factory village in which housing, medical care, education, and other amenities unknown to society at large were provided to workers in exchange for bondage, or what Ruth (1984: 78) terms "semi-subjugation." In the nineteenth century, this pattern of integrated rural economic development was given a new form as British mechanics helped develop canals, railways, mining, manufacturing, and modern weapons production in industrial workshops, textile mills, and other manufacturing communities, often located outside the large cities.

A third feature is the systematic and long-standing incorporation of expertise into the established social order, which has affected the ways in which intellectual life has been organized and institutionalized. In Sweden, to a much larger extent than elsewhere, intellectuals are structurally (even symbiotically) attached to the activities of state, industry, and, in our day, mass media; even the universities are part of a unitary national system under one central authority, which was originally church and monarch and which is now, of course, the state. As a result, the comparatively few "free-floating" and therewith oppositional intellectuals—for example August Strindberg and Jan Myrdal—were and are found almost exclusively in artistic and literary circles. Academics, who are first and foremost civil servants and subjects of the state, are often drawn upon as experts by the multitudinous "investigative commissions" that are established by central government to address new challenges. Again the pattern was set early on, during the great-power era. Lindroth (1952: 10) notes: "Higher educational life in Sweden during the seventeenth century, despite its gratifying vitality, was characterized by a certain one-sidedness of purpose. The prime task of the universities was to train priests and, to some degree, officials."

In 1866 the political tradition of interest harmonization was strengthened by a parliamentary reform that introduced a bicameral legislative system. This served to reinforce the political hegemony of conservative coalitions of landed aristocrats, well-to-do farmers, urban nationalists, and monarchist forces, including the central bureaucracy, which, toward the close of the century, was challenged by a growing labor movement and a modernizing liberal bourgeoisie (see Wittrock and Wagner 1992). This emerging tension among intellectuals between "traditionalists" and "modernists" came to color the twentieth-century debates about technology. However, with few exceptions, both traditionalists and modernists generally shared a common discursive framework, which assumed two basic postulates: that the consensual mode of governance was not viable and that Sweden should be reconstructed as a major power with technological and industrial attributes. Out of this framework grew the paternalistic idea of the *folkhem* (people's home), first introduced with a conservative connotation but later, especially after the onset of the Great Depression, given a social democratic interpretation as the model for a modern welfare state.

Still, it took the first four decades of the twentieth century to forge the modern, future-oriented sense of national identity in the name of "international solidarity" that is so characteristic of the Swedish model. The idea that Sweden had a unique moral mission in the world was, at least in part, born out of the debates about the implications of modern technological development. Sweden, the bastion of social engineering and the planned welfare state, could offer both exemplary public policies and financial assistance to the less fortunate nations. Particularly after World War II, when Sweden came to play an important role in the United Nations, this international modernism has been one of the main elements of Swedish foreign policy. It was perhaps most clearly personified by Dag Hammarskjöld, who served as the UN's secretary-general from 1953 until his mysterious death in Africa in 1962.

A significant indicator of the politics of harmony and consensus may be seen in the way in which Swedish employers and industrialists, even in response to the nationwide general strike of 1909, adopted a conciliatory attitude toward the rising labor movement. No doubt the fear of continued massive emigration to the United States played a role; but there was also the tradition of commissioning specialists to diagnose problems in the body politic and recommend solutions. Over the course of the twentieth century, the system of expert investigative commissions became more formal, expanded vastly, and served to draw many intellectuals into the processes of social, political, and economic reform. In other words, Swedish intellectual life was characterized by a pattern of discursive practices that favored conciliation, compromise, and accommodation across class barriers. This pattern is formative for the debates over technological development in the early twentieth century.

In 1909, Gustav Sundbärg, the secretary of a government commission set up to investigate the causes and consequences of emigration, tried to identify the fundamental traits of the Swedish national character. Sundbärg considered the mechanical interest of Swedes axiomatic; he attributed it to the climate, to the geography, and to the love of nature so characteristic of Swedish culture. For Sundbärg (1911: 9), Swedes were more interested in nature than in people, and they were thus more interested in the non-human machine than in human psychology. Sundbärg (ibid.: 11) compared how the typical Swedish boy and the typical Danish boy related to a machine:

Imagine, for example, a Swedish boy who is charged with maintaining a machine. After two weeks he will have mastered the entire construction and be coming up with improvements. If the machine breaks down, it would have to be extremely serious if the boy couldn't fix it himself. Now put a Danish boy in the same situation. At once he will feel a certain antagonism toward his machine, and it would never occur to him to devote any time to investigating its operation. If the machine breaks down, the boy would not be particularly bothered but would wait until a maintenance man could come and fix it.

Sundbärg's identification of Swedishness with mechanical interest and love of nature is by no means as ridiculous as his characterization of Danes. Indeed, it has become a part of the conventional wisdom among students of Swedish national identity to continue to stress similar features. This is, no doubt, due to the influence that Sundbärg's highly successful book has had on such discussions, but it may also be due to the importance of stressing such attributes every now and then in mobilizing popular support for the modern national welfare state. National identity, in other words, is itself a cultural construction (Löfgren 1993), and early in the twentieth century that identity was generally positive toward things mechanical, even imputing to the typical Swede a sympathetic relation to machinery.

Mechanization as such had few opponents in Sweden, which has experienced few if any Luddite revolts. Such controversies as there were concerned the consequences of technological development more than its actual content. The traditionalist rejection of modern technological civilization that was to be found in other European countries was quite weak in Sweden. Even the conservative poet Werner von Heidenstam kept a brand new automobile in his garage next to the rural mansion he built in the middle of Sweden, and Heidenstam found room for modern poetic meter as well as for modern technology while glorifying a mythical Swedish past (Kylhammar 1985). The younger postwar literary generation, with the exception of Harry Martinson, was unabashedly modernist and enamored with the technological wonders of the twentieth century. Even Martinson, however, as one of the "five young" poets, who published an influential anthology together in 1929, was not entirely negative toward the possibilities for freedom and mobility that were intrinsic to the new means of transportation and communication (Holm 1965; Kylhammar 1994). The "five young"—Erik Asklund, Josef Kjellgren, Artur Lundkvist, Gustav Sandgren, and Martinson—were the literary equivalent of the functionalist artists and craftsmen who would give Swedish industrial design and social engineering a well-deserved international reputation. Even the spirituality of Pär Lagerkvist was formulated in a lyric tone reminiscent of functionalist architecture: concise, pure, "streamlined."

In the great-power project, both in its conservative version and in its later, more hegemonic social democratic versions, several disciplinary discourses were mobilized, each reflecting the attitude of an expert intellectual more than a purely academic stance. According to Wittrock and Wagner (1992: 233), "at least three alternative discourses were continuously reproduced in institutions of higher education." These were the discourse of the new economists, the discourse of demographics and statistical research, and the discourse of the new sociology-oriented political science. Among the new economists were Knut Wicksell, David Davidson, Eli Heckscher, and Gustav Cassel, whose work "laid the foundations for the later prominence of the so-called Stockholm school which exerted decisive intellectual and political influence in the 1930s." Prominent in demographics and statistical research was Gustav Sundbärg, concerning whom Wittrock and Wagner (1992: 234) note the following: "Despite the deep conservatism of Sundbärg himself, he pinpointed the negative effect of the late expansion of Sweden's economic infrastructure and of delayed popular participation in governance." Practitioners of the new political science reinterpreted the classical discipline of *statsvetenskap* ("state science"), founded at Uppsala in the seventeenth century, early in the great-power era. They combined "contemporary history, constitutional theory and a broadly conceived sociological understanding of societal change" (ibid.). A leading figure in the field was Rudolf Kjellén, the first professor of government and geography at the new university college in Gothenburg, a "policy intellectual" who was also a conservative member of parliament and an active public debater. At Gothenburg, young conservative intellectuals tended to orient themselves toward Germany, while their colleagues at Uppsala were more inclined to identify with British conservative thought.

All three of these disciplinary discourses contributed centrally to the Swedish debate on technology. In one sense, there was never any real controversy. Almost all participants remained true to a basically instrumental and optimistic view of technological development. There was little if any of the existential antipathy that was expressed in Germany or the United States, and there were few attempts to escape from the wasteland of modern mechanical civilization to a pre-technological, classical existence. It was not technology itself that was at issue but rather its functions and consequences, its costs and its benefits. Although some would seek to conquer the machine and to tame technology in the name of some more balanced civilizational ideal, they were not so much opposed to technology as to the consequences of mechanization on the natural environment. The challenge of technological civilization was met with, at most, calls for an "ethical renaissance" under the inspiration of Albert Schweitzer, whose cultural philosophy was published in Swedish in 1924. The dominant tendency was to accept industrial machinery and technological rationality as crucial instruments in the reconstitution of Sweden as a great power (Björck 1992). Both acceptors and assimilators saw Sweden as having a unique role to play in the modern technological world as a leader in technological development and in the formation of what might be termed a modern technological aesthetic that would combine the best of the natural and the man-made in a new functionalism. Especially in architecture and industrial design, this functionalism would win Sweden a certain international reputation as early as the 1930s (Sandström 1989); it can still be seen at work in the furniture of IKEA and the artistic glassware of Orrefors.

If there was a great deal of agreement, there were nonetheless differences between the generations of intellectuals, due largely to differences in experience. The generation formed in the closing years of the nineteenth century experienced the coming of the modern society, partaking of the futuristic spirit that was dominant at the turn of the century as new voices, representing new social actors, competed for cultural hegemony with the privileged classes. The new generation, reaching maturity in the 1910s and the 1920s, was conditioned by a different set of factors: the experience of World War I and the revolution in Russia, on the one hand, and the establishment of new political and cultural elites on the other. In the 1920s, many of the modernists who before the war had been underdogs fighting for influence and power had become new figures of authority. It was the younger generation of intellectuals who questioned the prewar idea of progress in the wake of World War I and who thus could refine or reformulate the great-power project in ways that could take account of some of its earlier limitations. Yet, in comparison to other European countries, there was relatively little criticism in Sweden of the specific technological results of modern industry; that would come later, in the 1970s, and even then in rather muted form. One reason for this generally sympathetic attitude to technology on the part of so many Swedes is, no doubt, the fact that Sweden, almost alone among European countries, has been spared the firsthand experience of twentieth-century technological warfare.

The Political and Economic Context

Sweden's intellectuals responded to the technological challenges of the early twentieth century somewhat differently than intellectuals in the belligerent nations. Not having taken part in World War I, they were not so strongly affected by it; instead, they were able to recombine influences from the United States, Germany, Russia, Italy, France, and even India into a home-grown synthesis. What is striking about Swedish debates about technology is the comparative continuity between the 1910s and the 1920s. The watershed is 1905, when the union with Norway was officially broken and Swedish national pride was at a low ebb. For many intellectuals, Sweden's new position in the world could not be derived from military or geographical expansion; it had to be based on technological prowess. In the 1920s this vision, previously propagated mainly by conservative intellectuals, would be taken over by a new left influenced by the Russian Revolution. On the basis of modern technology, it was argued, Sweden should once again become a great power. The rational, mechanical Swedes could harness the technological wonders of modern civilization by reinventing the state bureaucracy as an instrument of technological modernization.

After World War I, Sweden emerged as an industrial power of growing significance. Industrialization had come late to Sweden, but in the last decades of the nineteenth century the country had caught up quickly. "Sweden's industrial development, indeed, probably occurred more swiftly than any other European country's at that time." (Jörberg 1970: 6) Innovations of the 1870s and the 1880s stimulated the expansion of several large export firms—L. M. Ericsson, SKF, ASEA, Alfa Laval, Nobel—which remain important on Sweden's economic and technological landscape today. As has already been noted, Sweden's industrialization process was based in large measure on engineering and on the work of engineers, but many of the key innovations took place in the traditionally rural-based industrial branches of forestry, mining and metallurgy, and mechanical construction. Particularly important was the development of the pulp and paper industry, which benefited from the harnessing of the hydroelectric power of the northern rivers. Pulp and paper, ball bearings, matches, steel, telephones, and machinery were not produced primarily for the small domestic market but were, early on, integrated into internationally active corporations. Thus, Sweden's industrialization was highly export-oriented. "Sweden's industrial development was in high degree a process of adaptation to events outside the country's frontiers. Only to a lesser extent was it an independent process of economic expansion." (Jörberg 1970: 80)

Swedish modernization did not bring about the same degree of urbanization found in Britain and later throughout the European continent; there were no large industrial centers comparable to Manchester and Lyon. The degree of polarization between social classes was also weaker in Sweden. The bulk of the population continued to live in rural settings and derive its livelihood from agricultural produce. City dwellers remained a minority until well into the twentieth century. In 1910, 75 percent of the population still lived in communities with fewer than 2000 inhabitants.

The social and economic consequences of industrialization for Swedish society were enormous, however. In a few decades, Sweden was transformed from a backward outpost on the periphery of Europe to a rapidly industrializing nation, undergoing in a short time processes that unfolded much more gradually in other countries. One of the results was a huge emigration to North America, particularly among the ever-more-redundant rural laborers. Roughly one-fourth of the population left Sweden between 1860 and 1910. There was political mobilization too. Social movements emerged to provide new collective identities for the displaced peasants. In the wake of industrialization, temperance, evangelical, and socialist movements developed and became important shapers of modernity within the Swedish political context. These "people's movements," as they have come to be called, organized the population in a constructive, reformist way, making it possible for Sweden to avoid some of the more violent social tensions that developed elsewhere in Europe (Lundqvist 1977).

Sweden's Social Democratic Party grew rapidly after its founding in 1889, eventually falling under the influence of Germany's reformist and parliamentary party. The general strike of 1909 marked the end of labor confrontationism in Sweden. By 1919, when Hjalmar Branting formed a coalition government, the Social Democrats had become professional politicians. In return, they were given a formative role to play in the shaping of the modern Swedish society.

The social democratic breakthrough at the end of World War I was a peaceful revolution that transformed a traditional and authoritarian society, ruled from above by a militaristic conservative elite, into a representative democracy with strong commercial ties to Britain and the United States and a concomitant political orientation toward Anglo-American liberalism. Ruth (1984: 80) has written: "Between 1909, the year of the general strike and 1918, the year of full male suffrage, a change took place—without bloodshed—far more dramatic than any that had occurred in other Western democracies in a comparable period. . . . Humanitarianism and internationalism replaced the nostalgic nationalism of former decades." Under social democratic influence, internationalism, and, with it, a future-oriented, functionalist nationalism, supplanted the backward-looking conservative nationalism that continued to dominate the political discourse in many other countries.

Although Sweden did not take part in World War I, it was not spared the economic turmoil that came with the war. The years 1917 and 1918 were the most difficult that the Swedish population had experienced since the 1860s. Until 1922 there was a further decline in industrial growth in the aftermath of the war. Food and other essential products were in short supply, and Sweden signed trade agreements with Britain, France, Italy, and the United States. This geopolitical reorientation, which was both commercial and diplomatic, soon became a target of attacks from the Swedish right, which decried the ebbing of Germany's influence. At the same time, important changes took place in national political arenas: suffrage was universally extended, and the right lost its dominance. In 1917 a leftist coalition government of liberals and social democrats was installed.

In the long run, however, the war had a positive effect in neutral Sweden. While the other European nations were busily destroying one another's industrial capabilities, Sweden's export economy was booming. In 1928 the value of Swedish exports was twice as high as it had been in 1913. At the same time, Swedish industry was actively taking part in the latest phase of technological development. The first Volvos were built in the 1920s, and radios and telephones diffused as rapidly in Sweden as anywhere else in Europe. By 1930 Sweden boasted more than half a million telephones, compared to 20,000 in 1890, and the telephone network had quadrupled in the same period (Kylhammar 1994). Perhaps the low population density and the dispersal of population centers encouraged the spreading of communication technologies in a much more substantial way than in other parts of Europe. In the mid 1930s there were even suggestions that the popular tradition of study circles and public lectures could be terminated and replaced by radio-communicated education. Although study circles have continued (indeed expanded), radio remains an important means of education in Sweden (Leander 1978: 362). Of course, the traditional Swedish bias toward things mechanical encouraged all this. The cultural assimilation of the new technologies was relatively unproblematic. Automobiles, radios, and telephones had few critics. Even in the most rural settings there was general appreciation of new techniques and machinery.

After World War I, while the younger generation of radical leftist intellectuals looked to the Soviet Union as the model modernist society, many conservatives, who once had looked to Germany for inspiration, looked to nationalist movements in France and Italy, and eventually to Nazism in Germany. Liberals and social democrats, on the other hand, were influenced mainly by the Fabian socialism of the Webbs and H. G. Wells and by the pragmatism and social reformism of John Dewey and Jane Addams. But there were also influences, on both the right and the left, from India. Rabindranath Tagore, who in 1913 had been the first non-Westerner to be awarded the Nobel Prize for literature, toured Sweden in 1921 (Björck 1992). And there was a receptive climate in Sweden for "alternative" ideas about technology that came from the Russian anarchist Kropotkin and from the British Arts and Crafts movement. Furthermore, some of the more moderate criticisms of technological society, such as that of Walther Rathenau in Germany and that of Stuart Chase in the United States, resonated in Sweden (Kylhammar 1990). It is perhaps not irrelevant that Sinclair Lewis, an American modernist, received the Nobel Prize for literature in 1930. For internationally oriented Sweden, so dependent on foreign markets and so open to foreign ideas, the war marked more of a turning point in terms of foreign influences than in terms of the dominant images of technology.

Because of these particular contextual influences, the intellectual debates about technology were generally couched in friendly terms, even though there were skeptics and critics on both the right and the left. Technological enthusiasm (focused on the achievements of Swedish engineers and entrepreneurs) became one of the central features of the Swedish public discourse. Ivar Krueger, the "match king," who utilized mass production and rationalization to amass a fortune and significant influence over the Swedish economy, became a national hero in the 1920s, incarnating—for both the reformist left and the right—the image of a new Sweden, a peaceful and modern great power combining technical ingenuity and commercial success. The 1920s were years of industrial concentration and of technical and organizational rationalization, scientific management and mechanization going hand in hand as Swedish industry leaped onto the international stage as a contender to be reckoned with (De Geer 1978). The Swedish rationalization movement translated Taylorism and American management philosophies into the Swedish, rural industrial context; in the 1930s, it would be carried over into a rationalization of both housing and family life in the social democratic policies of Alva Myrdal and others (Nilsson 1994).

The "Young Right"

Whereas in Germany a majority of academic intellectuals were critical or even hostile toward industrial development in the early part of the twentieth century, in Sweden—as we have already noted—few adopted this position. Furthermore, in most cases their critical stance only came after the war, and even then they were able to maintain a form of technological optimism that was increasingly combined with pessimist tones of cultural conservatism. Whereas the typical inclination of what Ringer (1987) calls the German-type academic mandarin can, with a few exceptions, hardly be found in Sweden, Jeffrey Herf's reactionary modernists did exist. (See Hård, chapter 3 in this volume.) Their shift to a more critical position after the war, moreover, was linked to a fear that large-scale industrialization was bringing about an American-style mass society and also to a fear that it would empower a labor movement influenced by socialist ideas. It was the October Revolution in Russia and the internationalist specter of Bolshevism, rather than any intrinsic antipathy to technology per se, that triggered this new wave of critique. An anglophobic tendency and strong identification with Germanic cultural values characterized the main opinion leaders on this side of the Swedish debate, the members of the "young right."

Early in the century the most vocal figures in this movement had written a number of ideological books and pamphlets that contributed to the resurgence of nationalist sentiments in Sweden. The titles are telling: *National Unification*; *A Program*; *Swedish Issues and Demands*; *Mass Culture*. The last of these was the most mandarin-like cultural and civilizational critique to be written in Sweden. Its author, Vitalis Norström, a professor of philosophy at Gothenburg College, was more a source of inspiration to the "young right" than an actual member. Norström's student Hjalmar Haralds also belonged to the group of Gothenburg conservatives. The hard core of this group consisted of the political scientist Kjellén and his protégé Adrian Molin. Nils Wohlin, an economist and a leading ideologist within the Swedish Farmers movement, was also closely affiliated. In Lund, Pontus Fahlbeck shared a similar perspective (Sörlin 1988: 196ff.).

These young conservatives were marginal to the main intellectual currents in Gothenburg, otherwise known for its strong anglophile liberal tradition, manifested in 1891 in the creation of the "free academy" (now the University of Göteborg) and again in the 1930s in an intellectual mobilization against Nazism. They were also outside the traditional conservative positions that were dominant in Stockholm and Uppsala. Their ties were, rather, with conservative intellectuals in Lund, and by extension, in Germany and Italy. Perhaps these ideological affiliations help explain the predilection for military metaphors in their propagandistic efforts on behalf of their vision of a new Sweden. It may also explain the transformation that took place in their vision as it was brought more into the mainstream of national politics in the 1920s and the 1930s.

Norström, perhaps because he was a philosopher, did not completely share the enthusiasm that the others felt for things industrial. He warned of the dangers of industrialization, which signified for him a kind of cultural and human degradation—a road to mass culture. Norström talked of *övercivilisation* (excessive civilization), which, he asserted, imposed a "new nature, which binds us to it like a straitjacket." There was, he claimed, "no other way to measure historical progress than through the growth of personality." It was the exaggerated nature of the technological project that he rejected, rather than technology itself (Björck 1992: 48–49). Kjellén, on the other hand, developed a concept of geopolitics in which industrialization played an important role. In his view, countries with rich natural resources and the capability to exploit them with modern methods had a comparative advantage in the global play for power among nations. "We great power Swedes (*storsvenskar*) dream of a Great Sweden within our own national boundaries, a new national greatness, but now on the strong ground of material production, with its enormous markets to the east, the old route of Swedish greatness." (Kjellén, *A Program*, quoted in Sörlin 1988: 197)

Molin elaborated on this conception in more propagandistic terms. After graduating from Gothenburg College in 1906, Molin went into newspaper journalism, and in the following year he started up a journal of his own. *Det Nya Sverige* (*The New Sweden*), published until 1928, was the voice of the young Swedish conservative movement, providing a nationwide forum for analysis, debate, and opinion.

Molin had good contacts with industrialists and engineers. His journal and the young conservative cult of the engineers promulgated a kind of technocratic ideology. Engineering was considered important not only for its central role in industrial development but also for the way it helped transform society and shape a new mentality with an emphasis on expertise, thereby playing down the role of politics—or, rather, turning political questions into technical ones. Kjellén (1908) claimed not only that engineering had this social side to it but also that it would make trade unions and the class-conscious workers' movement obsolete. Turning important social problems over to experts would make it possible to bridge class differences and to reach a common "objective" ground for mutual understanding and action. On the other side of the political spectrum, this idea came to guide the Social Democrats, who were able to put it into a more operational mode in the 1930s and onward. Still, experience in social engineering was not lacking, nor was a passion for new technologies. The young conservatives' passion was, at first, for industrial technology, but after the war it shifted to the agrarian sector and to a predilection for small-scale technology that prefigured E. F. Schumacher's *Small Is Beautiful.*

Molin's belief in expertise included "social engineering" of the kind he found himself involved in when he served on national investigatory committees organized by the state: the Emigration Committee in 1910, the Single Home Review Committee in 1911, and the National Housing Commission in 1912. In this capacity, Molin was what we today might call an organization man. Following Kjellén's cue, Molin (1909) wrote:

> On the riches of our forests and mines, and the power of our rivers, we shall build up our modern industry, which will provide people with jobs and a high standard of living for more than double the population of our country.... We are still a small people. We should become a large and strong people, as we once were, a people that can win the world's respect. And we would become a happy people—happy in work, in self reliance and in our faith in the future.

When he outlined his vision of a New Sweden, Molin often used military terminology. He spoke of a "campaign" and compared success in economic growth to victories on the battlefield. Since victory on this battlefield presupposed an adaptation to international modernization, the educational system had to be modified. The schools, he insisted, had "to train our youth, not as civil servants but as businessmen and other practical persons," and "commercial and technical education should be given priority" (ibid.). The "liberal school" was too intellectual for Molin, excessively oriented toward abstract or theoretical knowledge, and poorly rooted in reality.

Nils Wohlin gave vent to the same kind of patriotism when he spoke of the potential that lay bound up in the country's abundant natural resources. Hydroelectricity, Wohlin maintained, was the energy form of the future. Coal belonged to the past. England, Germany, and the United States had based their economic power and their industrial prowess on coal, but now, with the coming of the electrical era, countries like Sweden had a distinct advantage and could move to leadership positions. To what extent this would actually happen, Wohlin (1918) thought, would depend on how well the new great-power vision took root in society as a whole, and on who stood at the helm.

The role of technology in this often-repeated scenario was that of a revitalizing force. Rational and effective exploitation of Sweden's rich natural resource base and mobilization of scientific, engineering and other kinds of expertise were crucial. Faster industrialization, more effective exploitation of natural resources, and shorter times taken to bring products to an international market all hinged on this. All three of these factors were important components in a new geopolitics.

In some respects the rationale outlined by Kjellén, Molin, and Wohlin is similar to contemporary arguments in the technologically advanced nations that refer to high technology and strategic research as prime assets in the struggle over future markets. The difference here is that in turn-of-the-century Sweden this project was permeated by a much stronger ideology of nationalism and by great-power ambitions, and that its rhetoric was attributable in large part to the loss of Norway in 1905. The young conservatives distinguished themselves from the more traditional conservatives by looking ahead instead of languishing passively in the glory of past conquests. The new battlefield lay ahead—in the civilian sector, in science, industry, and trade. On this new international battlefield, Ivar Krueger was eventually to become "king," the modern incarnation of Gustavus Adolphus II. The older and more established Conservative Party was felt to be too tolerant of the Liberals.

The young conservatives also rejected the anti-modern and even anti-industrial brand of conservatism, with roots in the landed aristocracy, that tended to counterpose city and countryside. As Wohlin put it in 1915, the New Sweden was to be at once an agrarian *and* an industrial state. The two sectors would be in harmony in one great *folkhem* (Wohlin 1918: 119). There was no room for technological pessimism; modern techniques and electricity would bring almost exclusively positive changes. And, since the natural-resource base played such a basic role, the northern region of Sweden, Norrland, became a symbol of the future. Here was a parallel to the historical unfolding of a frontier in the United States and Canada. Among Molin's favorite phrases was "Sweden's America is in Norrland." Sörlin (1988: 97) writes:

The new image of Sweden was oriented toward the future and focused primarily on the nation's productive capacity. Nature was important, both aesthetically and as a resource. The importance attributed to nature in the national character can be seen to a significant extent in the art of the time. With its natural resources and its rapid and dynamic industrial development Norrland played an important role. In many respects, Norrland, both symbolically and literally, was an equivalent to the American frontier.

Wohlin became a leading figure in the Farmers' Party after the war, and during the 1920s he served as Minister of Trade (1923–24) and as Minister of Finance (1928–29) in coalition governments before taking on the post of Customs Director in 1930. His critiques of some of the

social consequences of industrialization became more and more pronounced during the interwar years; having studied in Germany before
the war, he had been strongly affected by the anti-civilizational viewpoints
there. In the 1920s, when a racist ideology entered into Swedish science
and politics, Wohlin often spoke of the farmers as being the most racially
pure Swedes. Like many of his colleagues of the "young right", Wohlin
was shaken by the war and perhaps even more so by the Bolshevik
revolution. In 1918 he wrote: "Large-scale industrialization's result is
sooner or later barbarism and barbarism's result is bolshevism. . . . An
uncontrolled industrialization leads directly to the dictatorship of the
proletariat." (Wohlin 1918: 599) Wohlin was opposed to large-scale
technological development, not to machinery itself.

After the war Molin too changed his position. No longer seeing the
United States as a model, he became critical of large-scale industrialization, preferring small-scale enterprises. To build up large industries, he
said, was like constructing a modern Tower of Babel. Thus, after the
war the young conservatives shifted the accent they put on technology.
Now they played up the virtues of electrifying the rural districts and
touted electrification as an adhesive for the country as a whole, while
heavy industry was seen as a threat. Norström's civilizational critique,
his counterposition of machine against soul and humanity, was never
accepted by Molin and Wohlin, but their common ground against certain
trends in industrialization and mass culture was broadened. They increasingly opposed the growth of the city at the expense of the countryside;
for Wohlin and Molin, the farmers represented Sweden's future. In this
respect, their position in the technology debate drew closer to that of
the more conservation-minded writers and scientists who were actively
propagating for a more efficient utilization of natural resources. The
dream of a new America in the northern forests had faded by the 1920s;
in its place had emerged the idea of a rural modernity, clean, electrified,
functional, and technologically advanced but rural in spirit. In that
new version of the great-power project, Sweden was to become a moral
example—"a cultural great power, from where ideas are spread over
the world, and whose judgments are respected in the larger nations"
(Wohlin 1918: 31).

The formulation of the rural project would draw on other traditions,
as well as on the nationalistic great-power ideas of the prewar "young
right." Particularly important were the attempts by a number of nature
lovers to bring an environmental consciousness into the Swedish political
culture.

Pastoral Visions: Forsslund, Martinson, Selander

In Sweden, with its Linnaean heritage and its vast forests, the early twentieth century witnessed the development of a conservation movement and the emergence of a strong tourist association. Swedish ecologists and nature lovers were, however, not directly concerned with technology. Theirs was not so much an opposition to the technological civilization as it was a movement to broaden or popularize the traditional love of nature. Here again, the typical Swedish ecological intellectual did not question or assess the technical instruments of modernity, as did at least some of his American counterparts. The negative sides of technology, which had become dreadfully apparent to intellectuals with wartime experience, were somehow less visible in Sweden. Technological development was largely taken for granted as an unproblematic feature of modern society; it would find few opponents until the economic crisis of the 1930s. As elsewhere, however, there were differences among Swedish nature lovers: between preservationists and conservationists, and between radicals and reformists.

The conservationists, who formed an association in 1909 and who succeeded in gaining official recognition for their efforts to establish nature reserves, contributed to the eventual appropriation of technology in Sweden. Like Nils Wohlin, they stressed the importance of spreading the spirit of modernity throughout the sparsely populated country; conservation in Sweden was as much a matter of encouraging the modern population to experience the natural environment as it was a matter of "protecting" nature from exploitation (Haraldsson 1987). But there was a tension between those conservationists who spent their lives in the countryside and those who were active in the early development of tourism. Among the Social Democrats were several notable socialists—Carl Lindhagen, Karl-Erik Forsslund, Paul Rosenius, Elin Wägner—who sought to infuse a rural spirit and what might be called an environmental consciousness into the Social Democratic movement. In this respect, the labor movement merged, in significant ways, with the farmers' movement; both developed their study centers and their recreational retreats in the countryside, and both combined a certain kind of pastoral rhetoric with their praise of modern technology.

The three representatives of pastoralism we will briefly consider here—Karl-Erik Forsslund, Harry Martinson, and Sten Selander—are indicative of the reach of the rural consciousness among Swedish intellectuals. All three were writers and nature lovers, but they identified with somewhat different movements and political orientations. Forsslund, who had studied the history of art and literature at Uppsala in the 1890s, was influenced by the anarchist vision of Kropotkin as well as by the aesthetic vision of the Swedish artist Carl Larsson (Sundin 1984). He became something of a utopian, dreaming of a new Sweden based in the countryside, with the aristocratic country estate transformed into a modern, egalitarian community. Though he wrote novels and poetry, he may be best known for his efforts to educate the working classes in the characteristically Swedish rural setting. Never an ideological socialist, he was one of the intellectuals active in the Social Democratic movement at the turn of the century who helped to plant the international movement in Swedish soil—for example, by making use of the Scandinavian institution of *folkhögskolor* (people's high schools), which had sprung up in Denmark and Sweden in the early nineteenth century to provide adult education for the rural population. Forsslund was active in establishing the Svenska Naturskyddsförening (Swedish Conservation Society), which in the early days was primarily an elite organization of scientists; he also helped launch the characteristically Swedish movement of *hembygdsrörelsen* (community preservation), which continues to educate foreign tourists and Swedish schoolchildren in the ways of Swedishness at various open-air museums that are to be found in the countryside and in most cities. He was, however, more of a preservationist than a conservationist, and he opposed the "relocation of the country into the city" (as at Skansen in Stockholm) with rebuilt village houses, traditional handicrafts on sale, and folk dancers entertaining a cosmopolitan urban public in colorful native costumes to the accompaniment of country fiddling.

Forsslund's vision of a decentralized, modern, authentically rural Sweden was to reappear in various guises throughout the twentieth century. Forsslund himself became a more detached nature lover after World War I, taking less of a part in political debates and social life and devoting himself to producing a 27-volume description of the landscape and culture of his beloved Dalarna, the quintessential Swedish region. Forsslund was most influential in the prewar period, when he saw in the new technological civilization both challenges and opportunities for updating the traditional Swedish value system—a system based on peasant virtues and intuitive natural understanding. He put it this way in one of his early novels, quoted on p. 352 of Sundin 1984:

The car is a wonderful means of transportation, the airplane is one of humanity's prettiest creations, and they certainly cannot be made at home by traditional means. Electric light, electric power are victories of the greatest kind. A power station, a factory can be a magic palace. But we have to make sure that it becomes one. And everything in its proper place! No motor on an old boat on Siljan [Dalarna's largest lake]—no motorboat within hearing distance of The Great Waterfall [Stora Sjöfallet].

Among the younger generation of writers and intellectuals who appeared on the scene in the 1920s, Harry Martinson came to be most closely identified with Forsslund's pastoral vision. It is no accident that Martinson was one of the early Swedish admirers of Lewis Mumford. Martinson was one of the very few writers in twentieth-century Sweden who explicitly challenged the technological imperative with an admiration of non-Western cultures and an almost Arcadian relation to nature. His nature poetry is among the finest ever written in any language. Never an organization man, Martinson liked to laugh at the technological hubris of modern man. He devoted half of a late poetry collection (Martinson 1960) to deriding the automobile as a dangerous toy that far too many adults liked to play with.

In the 1920s, Martinson wandered the world to escape the confines of technological civilization. With no homeland and no nation, but with an insatiable appetite for travel and freedom, he envisioned an international community of nomads. After returning to Sweden, he was one of the few writers to poke fun at the functionalism and technological enthusiasm of the 1930 Stockholm Exhibition. In an essay written in 1930 (quoted in Holm 1965: 34), he began to strike the tone that would reveal his characteristic attitude toward technological civilization:

We hear the singing of the grass and the lute playing of the cricket far out in the countryside where the grey porcupine crosses the road each night. But we hear another sound far away, a radio, a screeching whistle of a train. And we know that within a hundred years we will have to take a stand—for the City, pulled into it like a back-yard garden or accommodated to it: industrialized. But when there is no longer any room for the porcupine to cross the road . . . for the cricket to string his lute in the hay—then I do not want to live. Or I will be off to the inaccessible Territoria de Acre, far up the Amazon river—as long as they haven't built a boot factory there.

With the coming of the Depression and the rise of Nazism in Germany, Martinson grew ever more disillusioned. A trip to the Soviet Union in 1934 eliminated any hope he might have had that the future was being constructed across the Baltic in the land of Bolshevism. He increasingly left society behind, in his writing and in his life, to ponder the fundamental riddles of human existence from the perspective of a blade of grass or from within the "mind" of an insect. It was to biology, astronomy, and, eventually, an ecological holism of his own creation that Martinson moved in the 1930s. He did not so much criticize technology as he tried to avoid it, bemoaning the transformation of his beloved steamships into warships and the coming to power of the social functionalists whom he found so simple-minded (Kylhammar 1994: 163ff.). In later life, Martinson would become a kind of anti-modern prophet, a free-floating wise man; in poetry, he would question the modern technological civilization on behalf of a subdued and overwhelmed human spirit (cf. Sandelin 1989). The epic poem *Aniara* (1956), a modern (or post-modern) space odyssey would win Martinson a Nobel Prize in literature but little influence over Swedish attitudes toward technology and even less influence over technological development itself.

The example of Harry Martinson—increasingly extreme, detached, even anti-modern, but enormously popular among critics and readers—says something significant about the schizophrenic Swedish attitude toward modern technology. Though there was little room in the Swedish debate for a more balanced aesthetic critique of technology of the kind that flourished in other European countries, there was appreciation and understanding for a deeper and more fundamental questioning of modern civilization. Martinson's radicalism points to the relative absence in Sweden of an intellectual position that could correspond to that of Albert Schweitzer or to that of the Dane Poul Henningsen. But it also says something about Martinson's highly valued individuality: the radical critic of culture in Sweden—and in this respect Martinson connects back to Strindberg and ahead to Bergman—is opposed not merely to technological civilization but also to the state and, very often, to organized society. There is a certain primitivism in Swedish aesthetics that brings to mind the Viking heritage as well as the comparative difficulty that the individual artist faces in such a highly organized society. Martinson's popularity throughout his life, and his continued popularity in the years after his death, also indicates that his anti-modern, radical pastoralism continues to have a broad public appeal in Sweden, even though it has proved difficult to translate such a position into government policy.

A third variant of pastoralism, and perhaps one that has been more influential in the general political culture, is represented by the writer

Sten Selander, who has recently been remembered for a new generation in a biography by Martin Kylhammar. Like Martinson, Selander refused to "accept" modern technology on its own terms; through the 1930s, when he was most active as a literary critic, he criticized twentieth-century civilization's overdependence on machinery. Selander wrote the following (quoted in Kylhammar 1990: 43) in a 1930 newspaper article:

There must be something crazy with a civilization, which devotes more effort, intelligence and willpower to manufacture more or less necessary factory goods than to solve the fundamental problems of humanity's physical subsistence and spiritual health. The pulp factory at Ostrand stands as an impressive monument over human genius and creative ability; but at the same time the perpetual machinery in the empty halls represents a heavy sacrifice, the heaviest of all— the sacrifice of living persons.

Unlike Forsslund and Martinson, Selander did not try to escape from the technological civilization back to nature or into art, nor did he reject the technological civilization as totally as Martinson (at least in his poetry) did. Instead, as Kylhammar suggests, Selander tried to bridge the gap between the two cultures of humanism and engineering, and he did it in characteristic Swedish fashion, as an organization man, serving for many years as chairman of the Conservation Society. In that capacity, he managed to repair some of the damage that had been done by the relative elitists who had led the society in its early days. Selander did not want merely to protect nature from human intervention; he wanted to change the "general attitude about the relation between nature and people" (quoted in Kylhammer 1990: 115). In a sense, Selander carried Martinson's more artistic vision into the world of politics and organizations. In a book on Sweden's "living landscape" written just before his death, Selander summed up his position on technology:

We have all been taken over by a superstitious paralysis in relation to technology, which has become a superhuman Moloch, whose demands must be met whether they are reasonable or not. If you ask why, you get as an answer a slogan about raising the standards of living. But does it not mean anything for a high living standard, which also means a certain amount of comfort, how the country in which we live looks? Or is a high living standard only a multitude of cars racing through a broken-down landscape . . . ? (Selander 1955, quoted in Kylhammar 1990: 114–115)

These pastoral positions found a place, if a small one, in the Social Democrat-dominated society that began to take shape in the 1930s. But they were never really accepted; rather, in typically consensual fashion,

Selander and his Conservation Society were given responsibility for advising the government on conservation measures, and conservation was included (if marginally) in the reform programs of the 1930s and beyond. Kylhammar calls Selander one of the losers of the political debates of the 1930s, citing his defense of classical, bourgeois values and his attempts to argue for an integration of conservation and conservatism. But, in losing the battle, Selander formulated a critique of technological civilization that would be of some significance in the 1990s' transition to postmodernism. Selander, writes Kylhammar (1990: 210, 216), "formulated an ideology of conservation without falling into primitivism. He made it understandable how to protect nature for humanity's sake and which principles should form the basis for a humanistic conservation. Those problems—technology, the environment, humanism—were considered less important in the age of social functionalism than other problems—poverty, insecurity and unemployment. . . .There would be a long delay before Selander's concerns would dominate the political and cultural life."

Technocracy, Swedish Style

In Sweden, the positions that proved to be more influential for the social democratic synthesis of the 1930s were those that were more unabashedly positive toward technological development. There were, as in most countries, different kinds of technological optimists—those that were more elitist, or technocratic, in arguing for a new social and political role for engineers in the modern society; and those that were more "egalitarian" in emphasizing the value of modern technology for the common man and woman. In Sweden, it is somewhat more difficult to distinguish the two poles of the debate from each other, since the technocrats and the functionalists (as they might well be called) shared a great number of attitudes, both in regard to the state and in relation to political ideology. Because there is not much of a populist political tradition in Sweden, more active experiments with democratic control of technology such as have been developed in other countries have tended to be marginal there.

During the 1920s, while the technocrats and functionalists adopted a common attitude to technology, they nonetheless emphasized somewhat different aspects of "social engineering," and they tended to address different audiences. The technocratic pole was dominated by engineers and industrialists, who primarily worked within engineering and employers' organizations. Gerard de Geer, author of *Sweden's Second Great Power Era* (1928), was active as an industrial leader and as a public debater. Axel Enström, the longtime director of the Ingeniörsvetenskapsakademien (Academy of Engineering Sciences), was one of the main spokesmen for the new professional engineers. The engineering academy was established during World War I to raise the status and expand the social functions of engineers. As its leader, Enström helped develop a Swedish style of social engineering and industrial research that was to become increasingly important (Althin 1958).

The technocrats never really vied for political power; their aim was to raise the cultural status and the political influence of engineering science. In the 1910s and the 1920s, much of their energy was devoted to promoting various "rationalization" programs in industry and to promoting a national industrial policy (Runeby 1978). In the 1920s, some of the leading engineers felt called upon to defend modern technology from the criticisms that were leveled against it in the wake of the war. The criticisms were weak, and so were the defenses. Indeed, several scholars who have looked into the matter have asked whether it really makes sense to refer to Swedish technocrats at all. (See, e.g., Berner 1981 and Sandström 1989.) The kind of explicit technocratic vision that one finds in the writings of Walther Rathenau or in the later works of Thorstein Veblen, and in the activities of those who were inspired by those men, is absent from the Swedish debate. Instead, one finds the vision of Ludvig Nordström, who toured Sweden in the 1920s and 1930s as a journalist and who wrote influential books that, perhaps more than any other single contribution, paved the way for the welfare state and the Swedish model (Sörlin 1986; Kylhammar 1994). Nordström and the architects and designers who in 1930 sounded the call to accept modern technology were, in many ways, the true Swedish technocrats, and it was largely under their inspiration that the Social Democratic Party, upon coming to power in 1932, explicitly sought to constitute the "first modern nation" (Ruth 1984).

Social democratic inspiration, however, also came from outside influences. As in other countries, new groups of intellectuals emerged in the 1920s in opposition to the established political and economic order. In particular, the Soviet revolution encouraged the formation of a "new left" that combined Marxism-Leninism with more spiritual and vitalist tendencies. The journal *Clarté*, founded in the 1920s, propagated a modern radicalism in which left-wing social democrats could interact with more doctrinaire Marxists. Particularly notable here was the poet and philosopher Arnold Ljungdahl, who, in his long career, explicated Marxist theory for several generations of Swedish intellectuals. The Marxism that developed in Sweden, however, had little place for the critical attitudes toward technology that would enter into the Marxist philosophies of Korsch, Lukacs, Bloch, and Marcuse in Germany.

The functionalist position was especially strong in Sweden, less as an ideological glorification of the engineer and his particular form for technological rationality than as a popularization of instrumental aesthetics (Nilsson 1994). The aesthetic dimension of technology found proponents in the public debate, and it was perhaps most powerfully articulated by the aforementioned Ludvig Nordström. In a series of books and documentary films, Nordström attacked the dirt and ugliness of Swedish society and called for a regime of social engineers. Everything that was old and traditional, he argued, had to be eliminated. He glorified the new engineers who would rationalize the old, backward society, breaking down national borders and transcending class divisions (Kylhammar 1994).

The emerging Social Democratic leaders shared Nordström's technological optimism without necessarily sharing his "totalistic" political perspective. The pragmatic view was strong in the 1920s. In the early 1930s, Gunnar and Alva Myrdal brought the ideas of the Chicago School and the American progressive tradition to Sweden (Jackson 1990; Nilsson 1994). There was a great deal of American pragmatism in the Swedish model (as there would be a portion of Swedish social democracy in post-World War II American liberalism). However, that pragmatism was applied in a somewhat more all-encompassing way in Sweden than in the United States, since in Sweden the state and the government could exercise a much stronger influence over the national mentality. The forms of public enlightenment that had been developed in the social movements of the nineteenth century (the labor movement and the temperance and hygienic movements) were put to use in Sweden by social democratic ideologues. Alva Myrdal, through her involvement in women's organizations, played a particularly important role in diffusing the modern attitudes on matters related to the family, childhood, and education (Nilsson 1994), while her husband Gunnar helped form the economic policies of the Social Democratic Party. The characteristically Swedish forms of state intervention, involving rather large doses of socioeconomic planning, infrastructural support, active education, and research, and innovative measures for social welfare and job creation, owe much to the "rationalizing intellectuals" of the 1930s (Eyerman 1985; Wennström 1986).

After the Social Democrats came to power, they reached out to the Agrarian Party in an attempt to bring stability to the nation and to initiate a series of social reforms. Ernst Wigforss, longtime Minister of Finance, promulgated an even broader social contract that involved a kind of marriage between the state and private industry. In a letter to his brother in 1933, discussing the threat of Nazism, Wigforss (1954: 78) revealed the basic conviction that he would gradually develop into a set of policies: "I believe that social democracy among us [in Sweden] has a chance to capture people's imaginations and gather support for a strong socialist politics, if we could find a way to make a deal with the farmers. Without farmers and the unemployed, no real danger of Nazism." The longer-term vision beyond this immediate pact with the farmers was, Wigforss intimated, to effect a peaceful transformation of society so that the terms "bourgeois" and "socialist" would no longer have any meaning in Sweden. This was the same kind of consensus reasoning that informed Swedish functionalism, although for Wigforss there was more at stake than design and aesthetics. For him, the functionalist scenario had to move beyond the experts and the middle classes; it had to involve Sweden's industrialists. Wigforss encouraged state support of science and technical development in both new and traditional branches of industry as a way to broaden the basis of class cooperation. Apart from science and technology, the other areas that he considered crucial were natural resource management, foreign trade, finance policy, and the development of an infrastructure to help promote industrial production (Wigforss 1954: 117). In a speech given at Gothenburg in 1936, he used the term "reformist utopia" to characterize his vision.

Wigforss played an important role in developing technology policy. Having spent his student days as a Marxist, he saw technology and scientific-technical expertise as major ingredients in the modern democratic society. It was important, he felt, for scientists and expert intellectuals of all kinds to contribute to the collective project of social development. Speaking to students at Lund in 1937, Wigforss (1941: 181) said: "Engineers for material production, architects and planners for construction work, social scientists to fulfill the tasks of social engineering, and scientists in general will be not less but more in demand if we consciously direct our efforts to strengthen the basis of civilization and culture."

The key to the Swedish model was a historic compromise, and eventually a pact, between labor and capital. This was spelled out in the volume from which the foregoing quotation is taken, a book with the telling title *From Class Struggle to Cooperation.* The Social Democratic Party would hold power for almost half a century. In their long-term vision, science was an important productive force and a crucial ingredient for constructing the social machinery of the welfare state. Science and technology figured centrally in the pact between politicians and industrialists. As Wigforss (ibid.) put it: "If natural resources are one of the building blocks on which people build their welfare, discoveries, technical and organizational innovations are the other." In his capacity as Minister of Finance, Wigforss developed policies for the exploitation of natural resources, providing risk capital from the state in order to discover and utilize them. He also promoted the application of new technologies by introducing various institutional arrangements to allow private firms to benefit from state research institutes and laboratories without university interference (Elzinga 1993: 202ff.).

During and after World War II, Sweden developed a number of national research councils and branch institutes for cooperative research. In the 1930s, universities still lived in the world of the nineteenth century, professors still reigning supreme over their intellectual fiefdoms. The dominance of the professorial regime often extended to ideological loyalty. This being the situation, the Social Democrats did not see much need for state generosity in the traditional academic sector, especially when the main priority was to pull Sweden out of economic crisis. They were concerned mainly with science and engineering for industry, and, to some extent, medical science. Thus, for the universities the 1930s were a time of relative stagnation and isolation from the rest of society. In some cases, this reinforced their tendency to breed conservatism and reaction. After 1935, allocations to research institutes outside the universities began to exceed grants to universities. This moderated in the 1950s, but it was not until the formation of the doctrine of sectorial science policy in the late 1960s that the universities became the main repository of public research other than defense work (cf. Elzinga 1993).

The Swedish model that Wigforss and the Myrdals articulated in the 1930s did not really come into being until the 1940s, and it benefited from the neutral position that Sweden had taken during World War II. Whereas Sweden's productive capacities were not diminished by the war (quite the contrary), most of the other European countries had to rebuild their industries. The Swedish technocratic vision of the 1930s could thus also emerge relatively unscathed after the war. As with World War I, when Sweden had also stayed out of the fighting, World War II was generally not as traumatic for Swedish intellectuals as it was for other European intellectuals. The special synthesis of great-power conservatism and social democratic technocracy could thus withstand the multiple pressures that were exerted on similar projects in other countries. As a result, not only could the Swedish model provide a new kind of political consensus within the country; it could also exercise a certain attraction for other countries in Europe and other parts of the world in the years that followed World War II. By 1949, when the tensions of the Cold War were intensifying, Sweden's scientific prowess was considered significant enough by the US Central Intelligence Agency to warrant the evacuation of the country's leading scientists in the event of conflict. This is clearly stated in a recently declassified CIA document (Central Intelligence Agency 1949: 4):

It is clear that the totality of Swedish scientific effort will not seriously add or subtract from the US potential. Nevertheless, in view of the great capabilities of Swedish scientists, among them M. Siegbahn, A. Tiselius, and T. Svedberg . . . it would be of advantage to have them working for the US side. Probably nothing essential in the way of increased scientific contribution would be gained by forcing Sweden to join the Western Alliance, for already Swedish scientists collaborate well with the Western world. In the case of hostilities, however, it is felt that a considerable advantage would be gained by evacuating the top 20–40, or more, Swedish scientists to deny long-range benefits to the USSR.

Conclusions

What gives the various positions in the Swedish debate their coherence is what might be called a shared attitude toward culture. The "young right" sought to create a nationalist culture and a new sense of national identity based largely on the engineering achievements of the late nineteenth century. The pastoralists, on the other hand, were cultural spokesmen, trying to mobilize aspects of the Swedish cultural heritage— ruralism, love of nature, and even humanist or organic values—as a critical response to technological civilization. The idea of culture in this position is largely opposed to technology, but what is characteristic of Swedish cultural critique is its rich metaphorical and symbolic use of nature as a counterpoint to mechanical civilization. A rural rhetoric is the cornerstone of Swedish pastoralism: the city is contrasted with the countryside, the future with the past, and technology with nature. In a way, culture is reduced to nature, although both Martinson and Selander drew on other chords. But it is the sounds and the rhythms of nature that are stressed as critical alternatives to modernism. And in that respect, if no other, they too are taken into account in the social democratic synthesis. It is, as compared to other countries, a technological appropriation that remains sensitive to the "needs" of nature, and, indeed, oriented to the natural needs of humans (though not as much as the environmentalists, then and now, would like).

The victors, of course, were the technocrats and the functionalists, the direct sources of the Swedish model—the former representing an engineering culture and the latter representing an emergent democratic culture. It is the alliance between the two that is most characteristic of twentieth-century Sweden, and, as the twentieth century comes to an end, it is the alliance between technological rationality and egalitarianism that seems most challenged by new alliances of "green" businessmen. Now pastoral values and "cleaner" technologies are being promulgated by engineers and managers seeking to find new niches for Sweden in the global marketplace. Reactionary modernism seems to have reappeared in a new guise, and the future seems, in many ways, to promise not merely a replay of the technological debates of the early twentieth century but perhaps a turning of the tables.

7

National Strategies: The Gendered Appropriation of Household Technology
Catharina Landström

Technology, Ideology, and Housework

Many new technologies began to make their way into the household in the interwar years. Some (such as the vacuum cleaner) were said to simplify certain tasks; others (such as electricity) had the potential to effect more general or systemic changes. Some large-scale technological solutions even promised to make it possible to organize housework in novel ways. Although there was a striking unanimity about the advantages of the new technologies, they were, quite naturally, evaluated in different ways by different groups. Most commentators were clearly in favor of replacing or modifying traditional methods of housework by means of new technologies, but their reasons varied. Some groups argued that the mechanization of housework would allow social life to be organized in new, collective forms; others held that the new machinery would make it possible to raise the status of the traditional housewife. In other words, the new technologies were appropriated or domesticated (Lie and Sørensen 1996) in quite different ways.

In this chapter I will show how opinions on housework, on the role of women in modern society, and on technology were woven together into ideological positions, and how these positions linked technology to the "modern" home. I will also suggest how these positions were connected to debates on other contemporary issues.

In the United States, Germany, and Sweden there were competing views on how housework should be organized in modern society. At one level of analysis, representatives of different ideological positions advocated different technological devices as the best approach to particular tasks. At another level, advocates of competing ideologies attempted to explain and show how the new technologies fitted into social reality.

The connection between ideology and technology is a subject much discussed in social and cultural studies of technology (Pippin 1995). There seems to be agreement that ideologies play an important role in the introduction of new technology into society. Without implying any kind of functionalist or determinist understanding of the relationship between ideology and practice, my position is that ideology can be regarded as a necessary catalyst in the process of social decision making with respect to the stabilization of new technologies. If a new technology is to be accepted by users, it is not enough that it fits into familiar world views. Users must also be convinced that the technology will improve their lives.

In this chapter I will focus less on technology itself than on the ideas and ideologies that surround it. The multi-faceted term "ideology" is used to denote a coherent set of views and strategies that identify a problem, present a vision of how the problem is to be solved, and map out the most appropriate road or path to the solution. It is mainly the value commitments that unify and separate different viewpoints (Liedman 1984). The people discussed in this chapter were united by a concern over how the working conditions of housewives could be improved. The solution was defined by almost everyone in terms of technological improvement. Differences arose when it came to the definition of what kind of technology to push for and—even more strongly— when it came to what role women should play in appropriating technology into modern society.

"New household technology" is a vague term that can include electrification and gas lines, but I will use a narrower definition that refers to new electric devices designed to perform tasks of housework formerly done by human hands. These devices were central to the political agenda of the interwar period, and they were discussed in various publications. The discussions addressed practical questions, such as how various housework tasks could be performed and organized to make women's work easier: How can the cleaning of the home be made less time consuming? How can the dangers involved in cooking be reduced? How can doing the laundry be made less physically exhausting?

I rely mainly—but not exclusively—on recent studies of the interwar development and discussion of household technology. Obviously, this reliance poses a methodological problem, because if this body of literature produces and reproduces certain biases I will not be able to detect them. However, the benefits of using secondary sources outweigh the drawbacks. Without the massive groundwork done by scholars (see, e.g.,

Cowan 1983; Hagberg 1986; Koonz 1986), a cross-national comparison such as I attempt here would hardly be possible.

Besides the literature on household technology of the 1920s and the 1930s, my analysis is informed by the contemporary feminist discourse on women and technology (Wajcman 1991). This discourse, not explicitly discussed, is the foundation for my perspective, which is that writing the history of women as active creators of social reality is an important task.

The Discursive Framework of Home Economics

The theme that underlies this volume—how technologies are intellectually and conceptually incorporated into discursive frameworks of considerable strength and durability—runs through this chapter. This is why, in this section, I shall start with Home Economics as an international discourse that was formulated and reproduced in institutions and networks of individuals in several industrial countries. The Home Economics discourse provided reference points, and international communication was important for the similarities of the discussions in various countries.

Already in its emergent stage, the Home Economics movement struggled with the question of how best to adjust new household technology to housework practice. At the first international congress for Domestic Science, held in 1908, the International Federation for Home Economics was founded.[1] After the formal establishment of this association, international congresses were held regularly in the interwar period. The relationship between technological development and household work was an important theme of all these meetings.

The participants in the International Federation for Home Economics were mainly women from what Hagberg (1986) has called the "domestic establishment" in different countries. The domestic establishment consisted of groups, associations, and individuals that would hardly have considered themselves as members of a uniform group. But Hagberg argues that the many similarities in the social position of individuals, their socio-political ambitions, and their thoughts about the most important changes in the household make it reasonable to treat them analytically as a homogeneous group. Additional unifying factors were their having Home Economics as a basis for a full-time occupation and their social

1. For a historical overview of the development of the International Federation for Home Economics see *The Bulletin*, no. 2-3 (1972).

background in the educated middle class. They shared the ability to act as experts on household technology, consumer guidance, and Home Economics education. These influential people saw as their most important task to work for a modernization of housewives' conditions. Such semi-professional groups of experts on Home Economics existed in several Western countries, forming an extensive and important international network. Their analyses of the relation between the home and the rest of society could not be overlooked.

The international Home Economics congresses were held approximately every six years, in different locations and with different main themes. The first congress dealt mainly with reviewing the position of Domestic Science in different countries and determining this new academic discipline's relations to science and society. The second (Ghent, 1914) focused on how to teach children and adults good housekeeping. At the 1922 congress, in Paris, the aim was to develop methodologies of housekeeping education. The fourth congress (Rome, 1927) concerned the social role of Home Economics and the organization of housework. At this congress, Christine Frederick from the United States presented ideas about the rationalization of housework. Her program was modeled after Frederick Winslow Taylor's principles of "scientific management." The fifth congress (Berlin, 1934) was devoted to how Home Economics education could make use of various sciences. The sixth congress (Copenhagen, 1939) turned to how the intellectual, moral, social, and economic traditions of various countries affected the education of women.

The international congresses provided opportunities for the circulation of ideas and served to form a common discourse. The participants presented and took with them a variety of new ideas on the technological and social development of housework. Between the congresses, international communication flowed through *The Bulletin*, a journal published by the International Federation for Home Economics.

This forum for international contact and for less formal exchanges among individuals and groups certainly contributed to the formation of the three transnational ideologies of modernization I call *integration*, *industrialization*, and *collectivization*. The terms refer to three distinct ideological positions on the appropriation of the new household technology into modern society, which proposed technological, economic, and social reform respectively.

The first position, *integration*, entailed the idea that technological devices should be spread to every home without challenging the home and the family as traditional social institutions. The basic idea was that housework could be made less heavy and each task less time-consuming with the aid of new technology and through the rationalization of work procedures. The traditional tasks would still be done by women in the private home, but, because of technical progress they would now be easier to carry out and their quality would be improved. To an extent, this was the most technocratic ideology, since it avoided bringing social and economic changes into the discussion.

The second appropriation strategy, *industrialization*, focused more on the economic potential of the new technology. It was a call for large-scale technology to be utilized in industrial settings. The production of goods and services that used to take place in individual homes could now be transferred to small or medium-size businesses. The "modern home" would become a place exclusively for consumption and rest. Although united in their radical plea for the abolition of housework, the advocates of this ideology had no common view of what women should do instead of housework. For those on the left it was clear that women now had the opportunity to step out of the kitchen and form their own careers, but for those of more conservative leanings such a solution was not necessarily welcome.

Collectivization was, socio-politically, the most radical strategy. It was built around the idea that household technology should be adapted to collective use by the inhabitants of an apartment building or a community. The goal was to reorganize the social structure so that the private home no longer was the center for most housework tasks. Goods for consumption would be produced industrially, and the "service" tasks in the household would be done collectively. This was thought to reduce the amount of housework each woman had to do, and to give women time to take part in public life on the same terms as men.

In brief, the socio-technical visions of these ideologies can be described as follows: integration implied that household technology should consist of a range of small appliances in every home, industrialization implied that household technology should be replaced by industrial technology, and collectivization argued that household equipment should be utilized communally.

In the following I will examine these ideological positions as they were explicated in the United States, in Germany, and in Sweden.

The United States

Of the three countries discussed here, the actual introduction of new
household technology had gone furthest in the United States, regarding
both appliances and the infrastructure they required. During the 1920s
there was a rapid expansion in the spread of water conduits and sewer
pipes. For instance, 91 percent of the households in Zanesville, Ohio,
had water conduits in 1926, and 61 percent had both water and sewer
connections (Strasser 1982). According to the Chicago Department of
Public Welfare, only 7 percent of the poor had to use outdoor lavatories
in 1925. Electricity had been installed in 35 percent of all American
homes in 1920 and in 80 percent by 1941 (Cowan 1983: 93). During
the 1920s the prices of electric sewing machines, vacuum cleaners, and
electric washing machines fell markedly, and in the 1930s the electric
stove became a reasonable alternative to the gas stove. The most widespread electric household device was the iron: in 1941 79 percent of the
American households had acquired electric irons, 52 percent washing
machines, 52 percent refrigerators, and 47 percent vacuum cleaners
(ibid.: 94). The differences between the cities and the country were
significant, according to Cowan, as was the difference between different
wage groups. A well-off family in the 1920s might have a bathroom, a
gas or electric stove, a vacuum cleaner, and an electric iron, while the
worst off had no access to any of these appliances. In the 1930s a well-
off family might have an electric washing machine and a refrigerator; a
poor family had greater access to water and electricity, and perhaps a
bathroom.

During the period 1918–1939 the American debate was penetrated
by ideas of scientification and professionalization of housework (Bose
1979; Cowan 1983; Ehrenreich and English 1979; Fox 1990; Hayden
1981; Strasser 1982). The most dominant position in this debate was the
commitment to integration, which, to a large extent, was institutionally
tied to the academic subject of Home Economics (Brown 1985; Nerad
1987). The efforts for academization, which had begun in the nineteenth
century, began to bear fruit after the turn of the century. In the interwar
years, home economists arranged courses up to the college level to
educate housewives, servants, and teachers. Through the creation of
university departments, a foundation for professional status was laid.
The academically trained experts could teach housewives to rationalize
their work, make it more efficient, and raise its quality. Research departments would act as consumption guides by testing and evaluating new

goods for the home market—both consumer goods and technological appliances.

The home economists' belief in the blessings of technology was combined with ideas of rationalization. The goal was that households should function according to the same principles as an industrial facility. All household tasks were defined in production terms, described in great detail, and routinized in order that the need for personal knowledge and skill be minimized. Another goal was to do away with the notion of the home as a consumption unit (Matthaei 1982). A housewife's tasks were considered to be of the same productive kind as those of an industrialist, the argument ran, and the role of Home Economics was to help her to become an effective industrialist of the home.

Home economists in this period put forth arguments specifically aimed at technological matters. This side of housework had been thoroughly discussed at the beginning of the century within the Home Economics movement, and a sort of consensus had emerged. Virtually all home economists agreed that new technology might increase efficiency and quality. The evaluative framework had already been laid down. All that remained was to set objective test requirements and to institutionalize technological testing.

In the 1920s and the 1930s, "scientific knowledge," "professionalization," and "efficiency" were key words in American Home Economics, as in other areas of American culture. (See Jakobsen et al., chapter 5 in this volume.) The new household technology was welcomed, and it was regarded as a means of raising the quality of personal consumption, both quantitatively and qualitatively. The time savings achieved through technology and rationalization of working procedures would mean that the various tasks of the household could be done more effectively and better. One academic expert, Lillian Moller Gilbreth, summarized the integration ideology in the preface of a book:

Home-making is the finest job in the world, and it is the aim of this book to make it as interesting and satisfying as it is important. Waste of energy is the cause of drudgery in work of any kind. In industry the engineer and the psychologists, working together, have devised means of getting more done with less effort and fatigue and of making everything that is done more interesting. The worker not only spends his working hours more effectively and with more satisfaction, but has more time and more energy freed for other things. This book applies to the home the methods of eliminating waste that have been successful in industry. To the home-maker it offers a philosophy that will make it easy, and a method of approach that will make it interesting. The home belongs to everyone in it; each should have a part in both the work and the play. But on the home-maker falls the chief responsibility for success, and it is for her that the book is written. (Gilbreth 1929: vii)

Although the proponents of the integration ideology had the upper hand in the American debate, advocates of the other two views were not silent. What is interesting is that in the American setting the ideologies of industrialization and collectivization tended to converge. The main point was to get housework out of the private home and done either in paid businesses or in unpaid collectives. The industrialization advocates went furthest in viewing the private kitchen as outdated. Zona Gale wrote in 1933: "The private kitchen must go the way of the spinning wheel, of which it is the contemporary." (quoted on p. 17 of Hayden 1981) The solution to all the problems of housework would be to have household tasks performed by businesses owned and/or run by women. This line of argumentation—here expressed by Ethel Puffer Howes in 1923— was critical of technologies that were designed only for use in the private home:

Quite apart from the fact that millions of us are not able to command them, the washing machine won't collect and sort the laundry, or hang out the clothes; the mangle won't iron complicated articles; the dishwasher won't collect, scrape, and stack the dishes; the vacuum cleaner won't mop the floor or "clean up and put away." (quoted on p. 270 of Hayden 1981)

Howes belonged to the "industrializers," who wanted large-scale appliances to be utilized in a market economy. Small businesses, the "industrializers" asserted, should deliver the services needed in the household, and the new technology thus offered new business opportunities for industrious middle-class women.

Those oriented toward collectivization wanted small communities to utilize more or less the same large-scale appliances that the "industrializers" wanted, but on a non-profit basis. Some of the "collectivizers," grounded in the leftist utopian tradition, advocated reorganizing not only the family but the whole of society. The organization of housework was viewed as a constitutive part of the system that had to be changed. A fringe view of collectivization was held by certain religious groups (the Shakers, the Rappists, the Separatist Society of Zoar, and the Society of True Inspiration) that wanted to withdraw from society altogether. To these minority groups, alternative technologies for collective utilization were instruments for the creation of independence and self-sufficiency. Some of these sects even came up with their own technical solutions in order to maintain independence from the surrounding society. They

developed large-scale appliances not driven by electricity and not dependent on the steel and tool industries. Though their existence may not
have been unknown, such views were not loudly advocated in the discourse; they were used mainly as negative examples, for polemic purposes, by those who wanted larger-scale technology for social reasons.

Some had the economic resources to experiment with collectively
owned and utilized housework units run by hired personnel paid by the
users of the services. This can be compared to cooperatives of workers
and anarchist groups, who also managed to raise money for trying out
different forms of collective arrangements for living and housework.

Entrepreneurs in the service sector also showed interest in large-scale
housework technology, seeing a chance to establish themselves in lucra-
tive new niches involving the delivery of such goods and services as ready-
cooked meals, laundry, and child care. Their rationale differed strongly
from that of "materialist feminists," who saw the re-organization and
socialization of housework as the most important point in the project
of women's emancipation.

The wide range of reasons used to support large-scale household
technology proves the general point that there is seldom a functional
relation between views on technology and on the social organization of
work. At the one extreme was the view that the female members of a
group collaborate in making use of a certain technology as their duty
to God; at the other extreme was the vision that various household
tasks should be taken over by competing companies using the same
technology. Not surprisingly, the latter ideology became the only true
competitor to the integration ideology in the American debate. Slowly
but surely, collectivization became a marginal position.

Two reasons why the integration ideology gained the upper hand in
the American debate were that it was firmly anchored in the American
domestic establishment and that it had a strong institutional base (which
the other ideologies did not have). Home Economics and its rationalization branch, Scientific Homemaking, were developed by educated
women of the American white middle class. Owing to the fact that Home
Economics was an established academic discipline at this time, these
women often had a university degree in the subject. Most of the leaders,
including Ellen Swallow Richards, were college or university professors.
Home Economics was one of the rare subjects in which a female scientist
could have a career. Many women who wanted to do scientific research
ended up in Home Economics departments, where they became part
of a domestic establishment aimed at modernizing the American home

(Rossiter 1982). Both Lillian Moller Gilbreth and Christine Frederick had a foothold in Home Economics, and with the support they got from men they were able to establish themselves as experts on the rationalization of housework (Clarke 1973; Moore 1983). These experts received economic and practical support from various companies that produced goods to be consumed by private households.

The professionalized character of the American domestic establishment marks a difference from its Swedish counterpart, which, as will be elaborated below, was a movement made up of housewives and self-educated experts (Hagberg 1986). By comparison, the advocates of collective arrangements comprised a more diffuse collection of smaller groups, unable to summon the same influence in the public debate. Their ideas did not get beyond isolated experiments.

Germany

Germany was slower than the United States to embrace new household technology. Berlin had 95,000 customers for electricity in 1920, and this increased to 1.13 million by 1933 (Rieseberg 1988: 130).[2] With well over half of the households electrified, Berlin was nicknamed Electropolis (Langguth 1989: 44, 50, 103). However, electric stoves did not become common in Berlin until after World War II (Bussemer et al. 1988: 124), nor did washing machines or refrigerators (Langguth 1989: 105). The second most common electric item, after the iron, was the vacuum cleaner, which by 1929 could be found in more than one-fourth of the electrified households in Berlin.

As in the United States, the diffusion of new technology was fastest in Germany's middle and upper classes. Again, data on Berlin in 1929 give a clue. Among apartments with six or more rooms, 75 percent had irons and vacuum cleaners and 20 percent had an electric cooking arrangement. Among two-room apartments, 48 percent had irons, 15 percent vacuum cleaners, and 2 percent had electric cooking apparatus (ibid.).

All three of the transnational ideologies I identified above were active in Weimar Germany (Weismann 1989; Uhlig 1981; Tornieporth 1988; Koonz 1986; Bridenthal et al. 1984). As in the United States, advocates

2. As the largest city and the capital, Berlin was probably not representative of Germany. However, it was the center of cultural and political life, and thus the conditions there can be expected to have been important for the discussion.

of "integration" were most vocal. From the 1920s on, they advocated technification, rationalization, and the professionalization of housework, spreading their gospel in magazines specifically aimed at housewives. In the *Praktische Berlinerin* of 1924 one could read the following:

There is no area which has been more completely neglected by technology than the household—in particular the kitchen. Sure enough, we do have coffee grinders, sewing machines or potato graters, but they are—also in their latest designs—almost only made to be driven by human muscle power. Women know how much of their human power is consumed for never-ending, unproductive housework. (F. A. Seyfert, 1924, quoted on p. 170 of Weismann 1989)

If the new technology was to be spread throughout society, then houseworkers would also have to be the subject of extended education. In 1921 young women could read the following in *Die Frau*:

Most important would be a professional education aiming at giving each girl the pertinent knowledge and skill that prepare her for the double tasks of being housewife and mother. Only such an education will enable her to run a household in a rational manner and to utilize all the blessings of technical progress. (Wer 1921)

The theme of rationalization found physical expression in the rationally planned *Frankfurter Küche* (Frankfurt kitchen), designed on the basis of time-and-motion calculations and intended to minimize the number of steps a housewife had to take. This kitchen was exclusively a workplace, without room for social activities; it was extremely small (to save steps), and its shelves designed to minimize the number of movements. This idea of saving the housewife's labor went well with the need to keep building costs down, and so it was embraced in some of the building programs implemented by the Weimar government.

As in the United States, the integration ideology was mainly advocated by women belonging to the domestic establishment. In the early and mid 1920s their ideology of technology corresponded well with similar ideologies except for the emphasis they put on the notion of economic scarcity. (See Hård, chapter 3 above.) The hardships of war still affected the minds of the German women more strongly than the minds of women in other countries, and hyperinflation did not make things easier. These experiences, which underpinned the notion of scarcity, were turned into a drive to economize and introduce recycling, which in the integration rhetoric became an important reason for advocating technological appliances. The notion of scarcity also surfaced in discussions of the possibility of hiring domestic servants and in plans for the education of housewives.

Another distinct national feature was the discussion of housework's intellectual and cultural content, which gave rise to a heated debate over whether it should be viewed as work at all. The monthly *Die Frau*, published in Berlin from 1918 to 1933, was a forum for the middle-class version of this debate. Several questions about technology, economy, and the role of women in modern society were debated in connection with this theme (Lüders 1921).

The "collectivization" ideology was articulated more strongly by women in Weimar Germany than by women in the United States. One of the most debated questions concerned the *Einküchenhaus* (One-Kitchen House), the idea of which was that apartment buildings should be constructed with one central kitchen in which the food for all the inhabitants could be cooked, either in a collaborative effort by the residents or by persons hired for the job.[3] This central kitchen should be equipped with a dishwasher, a store room, and a washing machine. The central kitchen was thought to free middle-class women from a lot of heavy and time-consuming housework, and also to give back to them the valuable time that they had lost when the number of domestic servants fell.

The idea of a one-kitchen house, which had originated on the political left before World War I, was again taken up in the interwar years. One of the early advocates was Lily Braun of the Socialist Party. Her views were shared by some leftist women, but not all; Clara Zetkin, for one, argued that one-kitchen houses neither could nor should be built before a socialist revolution. In the 1920s the one-kitchen house gained popularity in the bourgeois women's movement. The idea was also propagated by some architects, female and male (Uhlig 1981). The Bauhaus movement's functionalist style typically entailed one-kitchen houses for middle-class people who had lost most of their servants, but it also entailed big apartment buildings with central kitchens for the workers. The *Einküchenhaus* was also taken up by technocrats who wanted a fully Taylorized society with all activities carried out in a rational and efficiency-optimizing fashion. Most of the ideas surfaced in the middle class and were realized in houses built for the same middle class, but there were also one-kitchen advocates within the workers' movement, where the idea was connected to a desire to change the social structure in the family and the community.

In the 1920s, when German architecture began to blossom and the Weimar government started a large housing program, the idea of the

3. Such a building was also to have central heating and a nursery.

one-kitchen house was revived. The Weimar housing program was aimed at building better houses for the working class, and the one-kitchen house was a way of saving both building costs and the energy of working-class women. This heavily state-subsidized reform program gave reform-eager architects the chance to try out their ideas. Each of these movements had its heyday during the early and mid 1920s; later they faded away, both as a consequence of Nazism and as a consequence of the difficulty of getting a one-kitchen house to work as intended. Still, any idea even close to collectivization was attacked by advocates of integration who feared that such arrangements would destroy the culture of traditional family life (Kempf 1923).

The nucleus of the domestic establishment of the Weimar Republic was the Reichsverband Deutscher Hausfrauenvereine (National Federation of German Housewives' Associations) (Bridenthal et al. 1984, Koonz 1986, Weismann 1989), formed in 1915, whose membership peaked in 1924 at 280,000. The RDH started out as a part of the bourgeois Bund Deutscher Frauen (Society of German Women). As it grew, it became more autonomous, and its bonds with BDF became less binding and more formal. The RDH consisted mainly of middle-class housewives who had been used to having servants but now saw their old privileges withering away. In the Weimar Republic both labor-friendly reforms and the widened labor market for women made domestic servants more of an upper-class privilege.

The Reichsverband Deutscher Hausfrauenvereine tried to counter the educational institutionalization of domestic skills with a program for the education of servants in the private home. This was a move aimed to weaken the domestic servants' union and strengthen the housewives' control over the individual servant. Apart from the regulation of the domestic labor market, the RDH also established itself as a body with technical expertise. It had representation in the Price Control Commission of the Prussian state, in the National Coal Council of the Labor Ministry, and in advisory commissions combating black marketing. It was also represented in local administrative organs, such as Price Control Commissions in the cities, Labor Bureaus, Consumer Chambers, and Agricultural Chambers. It got a lot of support from industry, which bought advertising space in its journal and supplied local organizations with samples of products for free or at cut rates for private use by members. In other words, the integration ideology had a strong institutional backing—just as in the United States.

The Reichsverband Deutscher Hausfrauenvereine differed from its

American and Swedish counterparts in that it increasingly took political stances during the interwar period. First, it acted as the employers' organization in relation to the organized domestic servants. Second, it dissociated itself from the women's movement and opposed the fight for civil equality. Third, it was expressly anti-left-wing. Fourth, it advocated an aggressive nationalism.

The politicization of the movement was a gradual process. In the beginning the Reichsverband Deutscher Hausfrauenvereine was very similar to its counterparts in other countries, but as time went by it became more overtly political, and by the mid 1920s the politicization took a turn to the far right. The argumentation became overtly anti-feminist, anti-left-wing, and nationalistic.

Until the Depression, the Reichsverband Deutscher Hausfrauenvereine pushed for rational, professional housekeeping, but with the depression came a new line: that the role of housewife was the only proper occupation for women. This distanced the RDH even more from the rest of the women's movement (Koonz 1984: 205). When the Nazis came to power, the RDH parted from its sister organizations to follow the new leaders, among other things opting to exclude all Jewish members from the association (Bridenthal et al. 1984: 166). There was general agreement in political views between the RDH and the Nazis, but there were some conflicts having to do with organizational jurisdiction and autonomy. Eventually, in 1934, the RDH was incorporated into the official Nazi women's organization, the NS-Frauenschaft (National Socialist Women's Group).

The NS-Frauenschaft/Deutsches Frauenwerk (National Socialist Women's Group/German Women's Front) was the only organization for women allowed in Germany after 1933. It was in harmony with the Nazi party (formally, Nationalsozialistische Deutsche Arbeiterpartei, or NSDAP), and its goal was to organize all German women in the mobilization of the country under NSDAP rule. The NS-Frauenschaft had been founded in 1931 with an initial membership of 20,000. In 1935 it was recognized as a section of the NSDAP. It was for party members only, while the Deutsche Frauenwerk organized all sympathizing women. Deutsche Frauenwerk was founded in 1933 as a mass organization to be led by NS-Frauenschaft. The NS-Frauenschaft had the same chairwoman from 1934 to its abolition at the end of the war: Reichsfrauführerin Frau Gertrud Scholtz-Klink. The hierarchically structured organization had several specialized departments, including one of Volkswirtschaft-

Hauswirtschaft (Economics and Home Economics). Of course, the Nazi way of organizing women excluded all groups with different opinions. The Nazis recognized housekeeping as an area important to the national economy (Tidl 1984). They brought housework into the interest sphere of the state (Koonz 1984; Müler 1979; Bridenthal et al. 1984). One important point in the Nazi ideology of housekeeping was to minimize consumption through economizing in the private housekeeping and through recycling goods. The argument for economizing made the marketing of new technologies quite complicated. On the one hand, most new devices required electricity or gas; on the other hand, the refrigerator would make it possible to store goods better, thereby reducing waste.

The Nazis wanted it both ways. They combined a positive attitude toward individual items with an advocacy of collective equipment. One of the reasons why this did not lead to a contradiction was their idea of the nature of women. According to the Nazis the two genders were divided in essence. Women were supposed to be active in three ideologically defined areas: motherhood, housewifery, and "feminine labor." At least four children were expected from every German woman, and the state rewarded mothers who had even more. The Nazi idea of housewifery held that it was the woman's duty to provide the German man with a home, a place were he could return after battle, and a place worth fighting for in war. In the home the housewife would do all the necessary work. Besides cooking and cleaning, she was expected to be able to do the necessary woodwork, needlework, gardening, and health care. Housework as defined in the Third Reich was viewed as a craft. According to a 1937 slogan, the housewife was to take care of the "5 K's: Kinder, Kammer, Küche, Keller, Kleider" (Children, Chamber, Kitchen, Cellar, Clothes) (Bridenthal et al. 1984). The third area for women, "feminine labor," was paid work that all German women and girls were assumed to be suited for: education, social work, nursery school teaching, cooking, nursing, teaching, dressmaking, laundering, and the like. To be a farmer's wife was also considered an occupation.

Because women were assumed to be suitable for these three areas, there was no need to discuss the consequences of different types of technology for the role of women in society. Women had their natural, biologically defined roles in society, which could not be altered, and those who did not accept their biological destiny could be severely punished (*Eldorado* 1984). Household technology could not change women's "natural" conditions.

Sweden

The new household technology was not particularly widespread in Sweden in the period under consideration, although a steady increase was discernible (Hagberg 1986: 33f.; Nyberg 1989: 55). Sweden was still largely an agrarian society, and the differences in degree of technification between countryside and city was very large. In 1920 electricity had been installed in 90 percent of city households, but only in 40–50 percent of the rural homes. However, by 1940 three-fourths of all households had electricity. Eighty percent of city homes and almost no country homes had water conduits and sewer pipes in 1920, but at the end of the 1930s 85 percent of city homes and 25 percent of rural homes had water and sewer connections.

Class also made a difference for the access to new technology. The working class had less access to the new technological facilities than the middle class, both in the cities and in the countryside. This difference applied to electricity, water, and sewage, as well as to household devices that could be purchased individually.

After the electric iron, the first technological appliance to become common in Swedish homes was the sewing machine (Nyberg 1989: 81–90). A 1941 study of its diffusion shows that 90 percent of all households owned a sewing machine, but most of them were manually operated. In 1933, 75 percent of the households in larger cities had stoves (mostly gas). In 1933, 80–90 percent of the city households had access to jointly used laundries, most without electric appliances. By 1941, as many as 89 percent of middle class households and 31 percent of working-class households had vacuum cleaners. Refrigerators were not installed in Swedish homes until after World War II.

Thus, most of the new technology discussed by the experts was not available to the majority of the households in Sweden. As in Germany, the debate concerning the consequences of the new household technology was well ahead of its actual introduction. This means that the debating groups often based their argumentation on quite weak empirical evidence, and instead made extensive use of scenarios and more or less commonly accepted views on the home, housework, and technology to gain attention and credibility. Nevertheless, the interest in discussing these issues was widespread, perhaps because it was felt that there still existed a real choice between different technologies and organizational forms.

All three transnational ideologies can be found in the Swedish debate, but just as in the United States and Germany the national context formed the debates and gave them a particular bent (Åkerman 1983a; Åström 1985; Hagberg 1986; Hultgren 1982; Kyle 1987; Nyberg 1989; Waldén 1990).

In the 1920s the debate was dominated by the Husmodersförbundet (Housewives' Society), which had been created in 1919. Its argumentation, propagated in the weekly magazines *Idun*[4] and *Husmodern*, clearly followed the "integration" line. At the outset the Swedish advocates concentrated on the replacement of human work by technological appliances. The Husmodersförbundet argued for the dispersion of electric irons, stoves, refrigerators, vacuum cleaners, and washing machines. According to spokespersons, these technological appliances would help Swedish women do housework more efficiently and with improved quality. The propaganda for these appliances often consisted of a rhetorical comparison of the performance of the new appliances with the traditional way of performing the tasks, such as the following:

Nobody with experience of the electric kitchen can avoid praising it. It is beyond doubt that the solution to the problems of tomorrow's kitchen to a large extent is presented by the electric kitchen and all its appliances. (*Idun* 1920).

It was also pointed out to the housewife how new technology would make her able to cope without the servants that formerly had been available:

After the Seves stove has been installed, the use of other electric household appliances has become cheaper. Electricity has replaced all three maids, and housewife Addi von Hofsten now manages the whole household without a single servant. (*Idun* 1925)

In the 1920s, inspired by the American Christine Frederick, the Husmodersförbundet also became interested in the organization of housework. This theme was presented in *Idun* in 1920 in a series of articles pointing out how much further the Americans had come with the improvement of housework. One article described Frederick's ideas at great length, and also presented a drawing that compared a rationally planned kitchen with an unplanned. The rationalization of work was presented as a good thing signifying progress:

Our time demands high speed and efficiency in all professional areas. The representatives of housework are obliged to take part in the progressive development of

4. *Idun. Illustrerad tidning för kvinnan och hemmet.* 1888–1963. Stockholm.

their profession. Also in this area there is a certain transformation needed in the direction of higher work efficiency—or, to use the modern term, rationalization. (*Idun* 1930)

The integration advocates also argued that women should be professionalized in their roles as housewives. This was a very important point, connected to the mechanization and rationalization of work (Lilliehöök 1930).

The integration ideology was not alone on the scene, however. In the early 1930s its advocates received strong opposition from parts of the Social Democratic Women's Association and some educated young men eager for the modernization of society, both groups arguing for the "industrialization" of housework. Their ideas were presented at the Stockholm Exhibition in 1930, where the spirit of functionalism, especially in architecture and building design, dominated. According to this alliance, the home should be equipped with new technology, but most production of commodities should be transplanted to industry in order for housework to be minimized. The family should be turned into a consumption unit with rest and recreation as its major functions. The working area of the home should be reduced to a minimum, and most of the space planned for rest, not for practical work. The few household tasks that still remained would be made less time-consuming with the assistance of modern technology. Although it was clear enough in its view of technology, this ideology became very vague when it came to what the former housewives should do with all their new spare time.

An interesting aspect of the Swedish case is that, rather soon, the industrialization ideas merged with those of collectivization—the latter appearing in the 1930s. The undisputed leader of the collectivization movement was the well-known politician and public figure Alva Myrdal, who represented a group of influential women in the Social Democratic Party. In the mirror of history Myrdal stands out as one of the most important people in the debate regarding the transformation of Swedish homes. (See, e.g., Bok 1987; Hirdman 1990; Lindholm 1990.) She was more academic than most other Social Democratic women. Together with her husband Gunnar she wrote articles and books on social conditions, in addition to a number of proposals for social reform policies. Alva Myrdal was already established as an authority on social issues in the 1930s. She was consulted by the political establishment to analyze urgent social questions, and she also introduced new subjects into the public policy arena. Most of the Social Democratic women allied with

Alva Myrdal in their view that housework should be carried out outside the home. Some came from the separate Women's Association, but most were engaged in party work or in the trade unions.

Alva Myrdal argued that the "housewife" was an anachronism representing a bygone stage in societal development. In modern Sweden men and women should participate in production outside the home on equal terms. This was to be accomplished by utilizing technology to collectivize the work that was formerly done in each home. This camp advocated large-scale technological appliances to be employed in collective kitchens, nurseries, laundries, and the like. The rhetorical cornerstone in this approach was the double call for modernization and the emancipation of women. Owing to Alva Myrdal's fame, the views could be publicized in women's magazines, daily newspapers, books, and various governmental publications. They were brilliantly argued, and included references to famous critics of modern society. Articulating a deep opposition to bourgeois ideals, Myrdal (1935) stated the following in a debate:

> If we apply Veblen's divining rod, then things take on a much clearer light. One conclusion seems to be unavoidable: if we do have a leisure class in our country, then it is to be found among the married women of the upper and middle classes. This group might not be particularly large, but it is growing. It is one of the first signs of an increased standard of living, when men from the middle and working classes try to buy their wives out of paid labor.

In *Idun* one can find both criticism and approval of these ideas, and during the period there are interviews and articles from people with conflicting views. (See, e.g., Kaj 1930; von Kraemer 1930; Jan 1935.)

Let us look a bit closer at the people who were advocating these ideologies. The integration organization, the Husmodersförbundet , had been formed in 1919 (Åkerman 1983b; Hagberg 1986; Hultgren 1982). Its origins had been in the governmental and private housekeeping activities introduced during World War I, and in 1929 the association had about 10,000 members. The members of the organization belonged to two social categories. One of them was what can also be viewed as the constitutive part of the Swedish "domestic establishment" in the period. Here we find educated women with a base in public society, belonging to the upper and the middle classes. Generally, they were not full-time housewives like the second category of women active in the different local branches of the National Society. Most of the members lived in smaller towns and belonged to the lower middle class. The social position of the front-line members is important for understanding the organization's shifting focus of interest. During the 1920s the efforts were aimed at replacing human power with machines to compensate for the reduced number of servants in private homes. In the 1930s the interest turned toward recognizing the qualified nature of housework and reevaluating the role of the housewife. In the same decade the improvement of education for housekeepers became a major concern. The Husmodersförbundet worked closely together with male-dominated organizations that were prominent in the contemporary debate on the modernization of the industrial part of the Swedish society, including the Teknologföreningen (Society of Engineers) and the Industriförbundet (Federation of Industries).

The social foundation of the industrialization ideology subsequently subsumed by collectivization ideas was, in the first place, the Social Democratic Women's Association, formed in 1920, most members of which were the unemployed wives of male members in the Social Democratic Party (Carlsson 1986; Hirdman 1983, 1990). The argument for a separate organization of women was that women should learn political work on their own before they were mature enough to participate in party work. However, a majority of the female members of the party were active in the unions and not engaged in the Women's Association. The latter group in 1937 had 22,000 members, most of them full-time housewives, while the number of female party members was as high as 70,000.

The men who propagated social planning formed a separate group, exchanging ideas among its members and acting as a collective unit in public debate (Hagberg 1986: 120ff.). Most members of the group, including Uno Åhrén, Sven Markelius, David Blomberg, Sigurd Lewerentz, and Ture Ryberg, were ambitious young architects. They took great interest in social planning on a large scale. Household technology and living standards were to them an important part of the making of the modern industrial society. Being academically trained, they took an active part in the debate on modernization in newspapers and literature. This group organized the Stockholm Exhibition in 1930. Much of their inspiration came from the German Bauhaus movement and from functionalism. Their social planning ambitions and their social reformatory aims brought them close to Alva Myrdal, who stressed collectivization. One concrete result of the merger of the industrialization and collectivization ideologies was the erection of several apartment buildings with very small individual kitchens, centralized facilities, and staff paid for by increased rent.

National Differences and Similarities

If we compare the debates on the new household technology along national lines, we find, not unexpectedly, both differences and similarities. The most striking similarities are unanimous embrace of the new technology and discomfort with contemporary conditions. Groups that opposed modern household technology were very marginal.

Let me begin with the differences. In Sweden the "collectivization" ideology, driven by Alva Myrdal and the Social Democrat women from the unions, gained much more attention and political influence than in either the United States or Germany. No doubt this was an effect of the political situation in Sweden, where, especially after 1932, the Social Democratic Party consciously began to build a power base. A very important political *Leitbild* (Dierkes et al. 1996) that was not limited to the Social Democrats was the idea of the *Folkhem* (people's home). (See Elzinga et al., chapter 6 above.) *Folkhemmet* was a vision of a modern society where all members had equal access to the necessities of life. Full employment and a decent place to live were important ingredients in the political program. Social reforms and the creation of a feeling of community among the people was the recipe for this goal. In this political climate, Alva Myrdal's vision of the home and of women's role in society was a perfect fit. Since the Husmodersförbundet was viewed as bourgeois by the Social Democrats, and the reality of full-time housewives was a remote world for the low-paid women in the unions, the Husmodersförbundet was viewed with distrust. The bourgeois ideal of the home was not a real alternative for a majority of working women in the cities and the countryside. Their lives involved the problem of a double workday, and that problem would hardly be solved by technological appliances they did not even have the money to buy. Among Sweden, the United States, and Germany, it was only in Sweden that the argument was put forth that it should be the responsibility of the state to effectuate the program of large-scale housework. It was also only in Sweden that a party with a leftist orientation gained real political influence that could counter the interests of industry in these matters. The collectivization position influenced the views of the Swedish domestic establishment, which did not push for hierarchical organization of a profession as its American counterpart did.

The public debate in the United States was clearly dominated by representatives of the integration ideology, who regarded housework as a task for women and pleaded for a technology designed for individual

use. Owing to its high degree of institutionalization, the domestic establishment in the United States had a very special position. Its members
exerted a strong influence on Home Economics education and gave
advice to housewives, but they did not get involved in political matters—
even in cases where these affected everyday life. The dominance of
integration ideas can be interpreted as an outcome of the weak social
reformist ambitions on the American political scene. The home was
viewed as an arena of individual choice, and not as an area in need of
governmental effort. An industry pushing for increased production, and
thereby for indirectly increased consumption, was in perfect line with
this political culture. There was no real interest in socialist-inspired
solutions to social problems.

The counter-ideology that seems to have gained some small attention
in the United States was industrialization. The idea of a market of household services furnished by small and mid-size businesses did not need
to be connected to any idea of social change, but could be coupled with
the consumerist ideals prevalent in the period and with the ideal of the
entrepreneur.

The German scene was split, but the political confusion and the economic problems in the Weimar Republic did not nourish the advocacy
of social reforms of the type discussed in Sweden. In the material I
have gone through I have not found any German advocates for the
industrialization ideas, and it seems that the only alternative to integration was collectivization. Because the founders of the *Einküchenhaus* idea
had been leftist utopians concerned with social change, this ideology
was marginalized as conservative views spread. The German domestic
establishment seems to have gained more direct political influence than
its counterparts in any of the other countries. This could be an effect
of their articulation of political views that were gaining in influence. The
expanding right-wing views, which held *Kultur* very high and embraced a
cult of masculine values of war and blood, did not promote anything
that would change the role of women in society (Hård, chapter 3 above).
The right was positive toward new technology as long as it was used to
preserve the German culture and its masculine values, not to change
social relations but to restore the greatness of Germany (Herf 1984).
The increased influence of this social philosophy paved the way for
Nazism and excluded any opinions on housework technology that called
for changes in social relations.

As I have already pointed out, one striking similarity in the national
debates is the lack of criticism of the new technology. Everybody was

critical of the ideas of opponents, but very few were critical of new technology as such. The criticism that was actually voiced was not aimed particularly at the new household technology, but more toward the marginalization of the home in the industrial society and the threat posed against the more spiritual values of the home. In Sweden it was argued that moderation of technological development was necessary in order to preserve the immaterial parts of life in the home. The approaches to homework presented by the technology optimists were seen by some as ways of reducing the housewife to a work machine and letting the material goods be the only surviving part of life in the home. Elisabeth Waern-Bugge (quoted in Hagberg 1986: 113) voiced worry about the degradation of the housewife in 1924:

The task of the housewife is not to be merely the working machine of the home. She shall be the center of the home, the stable point in the being of the home, around which everything else revolves. She shall be the sun from which the whole planetary system of the home gathers power and warmth.

Another critic, Elin Wägner, acted in the press as a critical voice with a sharp pen and a wide circle of readers. As a well-established journalist and author with a critical view of society and a background in the women's suffrage movement, Wägner developed an interest in "qualitative female values" (Lindholm 1990). Her reservations about the direction of the development of household technology and housework were grounded in the view that the control over the home was being removed from the housewife to industry and was falling into the hands of manufacturers of the technological appliances and goods. This implied that the female values were being forced from their last stronghold: the home.[5]

Also in Germany there were worries about the devaluation of the housewife, coupled with a negative attitude toward "Americanization" (Jakobsen et al., chapter 5 above). Integration proponents countered the first opinion with a claim that the technification of the home was no threat to the immaterial values of housework (Weismann 1989). The fear of Americanization was met by a rhetoric that claimed that the rationalization of housework could well be done in a truly German way and need not copy American conditions. The fully rationalized American kitchen was sometimes used as a negative contrast to show how the feeling for the family was at the center of German technological modernization.

5. See Elin Wägner's classic book *Väckarklocka* (1942), which criticizes the whole of modern industrial society.

The lack of criticism can be explained by several circumstances, but two factors seem especially important. First, the few critical voices in the almost univocal positive appraisal of the new technology and rationalization of housework were much less united than the groups pushing for new technology. They acted mostly as individuals, and they had no organized network of sympathizers to refer to. This distinguished them from the well-organized, industry-backed domestic establishment. Second, and perhaps more speculatively, the lack of strong critique may be explained by the observation that in the technical modernization of housework it was not possible to identify any group that would lose. Housework was a reality for all women, and all women would like it to be easier; the only question concerned which type of technology would provide most relief. Most advocates of new technologies presented them as completely unthreatening, suggesting that nothing except the housewife's workload and the amount of drudgery she faced would change with the introduction of new appliances.

New Household Technology in Gendered Realities

This chapter has indicated that the industrialized Western world is quite uniform when it comes to cultural and social questions concerning the individual and his or her well-being. This is supported by the fact that, except for Nazi Germany, all ideological camps argued in terms of what the individual woman wanted and needed in a modern industrial society. This implies that the debate on household technology was very dependent on ideas on women's place in society. All the participants in this discussion on household technology had to have some opinion in this question. The ideologies in Sweden and the United States showed coherence in that the groups advocating individually applied technological appliances were conservative in their views on what "the home" and women's place in modern society should be, while the groups arguing for large-scale technology wanted change in the established conceptions often in a radical fashion. This also applies to Weimar but not to Nazi Germany. The Nazis' social policies differed so greatly from those of the Western democracies that there was no contradiction in moving housework to collective units, using individual technological devices, and seeing women as producers of children and a home for the soldier and as a labor force in "feminine" areas.

Household technology is one of the areas where it is most clear that men create technology to be used by women (Cockburn and Ormond

1993). Since male engineers and designers tended to have different frames of reference than the female users, mediation was necessary between an area culturally defined as female (the home) and an area culturally defined as male (technology) (Waldén 1990). In the interwar years housework was an area viewed as exclusively female, by both women and men. Many women distrusted the new appliances because they were invented by men who had no experience of performing housework. The women questioned whether men could know how to improve something they never had known how to do in the first place. The situation of distrust eased when the women in the domestic establishment took on the task of testing, evaluating, and recommending new technology. Relying on their own experience of housework, they used a language that, stripped of technical expressions, accentuated the performance of the tasks, not the technical properties of the equipment. Unlike the advocates of large-scale technology, the domestic establishment wanted no change in social relations in the household, a fact that made it easier to get the attention of a lot of women (and men) who did not want to change their everyday habits but just wanted to raise their standard of living. In comparison, the proponents of large-scale technology put forward demands that industry was not particularly interested in promoting. They could also only assert their views with the support of political authorities, and they required that people should radically change their everyday habits.

One interesting point when speculating about common cultural conservatism is the fact that virtually none of the established debaters raised the demand that men should take part in housework. The idea actually surfaced once in a reader's letter to a German journal, but it went largely unnoticed. It seems that such an idea was far too radical even to be discussed seriously. Perhaps such a demand of altering gender relations was seen as threatening also by the women pursuing the public debate on household technology. Their very expertise was rooted in the notion that, as women, they had natural and unquestionable authority in the area of housework.

Another important reason ought to be the male-dominated administration. As long as the debate on household technology was seen as an area for women, the conflicts were not overwhelming, and the expert women could utilize their knowledge and feel useful to all women. In the aftermath of suffrage struggles, it was probably comforting for the political establishment to find women focusing on housework and caretaking. A demand for men to take part in housework could have created a collision with men, and with women who believed that their only area of control would be erased.

The confinement of housework to women, regardless of the preferred type of technology, can be viewed as a discursive constraint in the discussion. The reproduction of gender in this debate did not cross the line to demand that men and women share responsibility for the housework. However, the idea of femininity was still reshaped. Most obvious are the articulation of women's competence and the feminization of technology. All women discussing the issues viewed women as competent in handling technology, organizing work, and caring for the needs of others. For those advocating that women should work outside the home, these qualities were deemed important for the public sphere too, while those arguing for women staying in the home used them to ground a demand that society pay more respect to housewives. The femininity constructed in this discourse lay very far from the idea of the weak, irrational woman who had to be taken care of and be protected by a wage-earning husband—the core idea of the nineteenth century's bourgeois wifehood ideology. It was also far from the view of women as victims of technological progress. In the discourse on housework, women were depicted as in control. The latter view indicates the process of feminizing technology. Though femininity and technology are often constructed as opposites, such was not the case in this discourse. When women in the interwar period discussed the rationalization and technification of housework, they constructed women as rational actors in control of the latest new technology and organizing the interaction between machines and humans. In this debate women were constructed as technically competent from the outset. To a certain extent, they realized something that scholars of science, technology, and society have only recently began to discover: that there is a female discourse on technology which depicts women as active users, and not as passive victims (Cockburn and Fürst-Dilic 1994; Wajcman 1991).

Perhaps the construction of women in control of domestic technology also feminized this technology and made it less interesting to the men in control of the majority of technology debates. A technology for women was deemed largely irrelevant for intellectual discussion in the contemporary discourse, as it is today in historical inquiry by male intellectuals viewing themselves as representing humanity.

8

Dutch Conflicts: The Intellectual and Practical Appropriation of a Foreign Technology

Dick van Lente

An advocate from Dordt,
A brewer, a notary,
A foreign potentate,
A stock broker—
They felt a great deal
For the Rotterdam harbor,
Which in their recklessness
They had almost destroyed.

Thus ran the second verse of a protest song called "The Bread Robber" that circulated among dockworkers in Rotterdam in 1905.[1] The subject was the recently introduced *graan elevator*,[2] a machine for pumping grain out of seagoing ships into barges called *lighters*. The song used the machine as an image of the employers who had introduced it: hard and cold, sucking the lifeblood out of the workers even as it sucked the grain out of the ship, and yet found in the most "civilized" countries. The introduction of two of these unloaders in 1905 provoked a strike that paralyzed the grain trade in Rotterdam's harbor. As a consequence, the Graan Elevator Maatschappij (Grain Unloader Company) was forced to idle the machines. Two years later, when the company put them to work again, there was another strike, but this time the company won. More unloaders were introduced, and many jobs were lost.

Though a local affair, the conflict over the grain unloaders became a national issue that was hotly debated in newspapers and periodicals.

1. Gemeente-archief Rotterdam, Politie-archief, no. 3354, map 1905, N 7060. It was probably this song that was sung at meetings of the strikers; see *De Maasbode*, November 14, 1905, and *Patrimonium*, May 9, 1907.

2. Since the English term "grain elevator" usually designates a building for storing grain, I will use the term "grain unloader."

Most prominent politicians and union leaders took sides. The *elevator-kwestie* (elevator question) can thus be used as a kind of touchstone for the debate about "the machine" and its social implications that had been taking shape from around 1820.

By the end of the nineteenth century, a virtual consensus had been reached among the leading spokesmen of political parties and interest groups, in which technological innovation was acclaimed as a progressive force while problems associated with it, such as child labor and accidents involving machinery, were attributed to deficiencies in social institutions and to laws that could be amended by legislative and technical means. Resistance to modern technology was generally condemned by the leaders of public opinion as reactionary.

The *elevator-kwestie* shook this consensus. Thousands of workers were prepared to risk their jobs and suffer hunger in order to prevent the introduction of the machine. Most politicians and union leaders condemned the strike and stuck to a broadly optimistic view of machinery. But some Rotterdam labor leaders were not quite so convinced of the beneficial effects of technological progress, and they supported the strike. In passionate debates with their superiors, they criticized the standard interpretations of mechanization as superficial and impractical, and they proposed better analyses and better guides to action. The crisis provoked by the introduction of the unloaders exposed, as crises often do, strains and contradictions in the dominant ideology that had been allowed to slumber in quieter times. This makes the *elevator-kwestie* an ideal case for studying the role of ideology in the appropriation of, and in the resistance to, modern technology.

One may regard ideology as a means of putting otherwise disconnected experiences and perceived problems into a meaningful pattern. For example, most people have no opinion about "technology" and its problems in general, but they love their car, hate smog, and are exasperated by the breakdown of their videocassette recorder. A theory or an ideology of technology may explain how these experiences hang together, for example by showing the increasing human dependence on technological systems in the process of modernization. In the case of the grain unloader, it is possible—thanks to an abundance of source material—to reconstruct the extent to which ideological preconceptions that had been formed in the course of the nineteenth century could perform this meaning-giving function for the various groups involved in the conflict—entrepreneurs, politicians, labor union leaders, and workers—and examine what this meant for their decisions and actions.

The object of this chapter is, therefore, to analyze the practical meaning of ideology in one fierce conflict over technological innovation. I will start with a brief discussion of the intellectual appropriation of industrial technology by the different ideological movements in the Netherlands during the second half of the nineteenth century. Then I will introduce the main participants in this particular conflict and describe their role in the grain trade and the way the unloaders were to affect their lives. Finally, having set the stage and introduced the actors, including the iron monster, I will let the drama unfold and address the relationships between the ideas of the groups involved and their actions.

The Intellectual Appropriation of Modern Technology in the Netherlands from about 1850

During the second half of the nineteenth century, ideas about the role of technology in society were usually expressed as part of the great debate, carried on in all Western European countries, about the social problems that accompanied industrialization (van Lente 1988, 1992). Large-scale industry came relatively late to the Netherlands and developed rather slowly there. This started around 1850; however, owing in large part to the depression of 1873–1895, it did not gather much of a pace until the mid 1890s (Lintsen et al. 1992–1995). Discussions in the Netherlands about industrial society were therefore mainly based on observations about Britain and Germany. At first these discussions were dominated by liberals, who had become the most powerful group in the Netherlands after 1848; however, from about 1860 on, liberalism was challenged from several sides. Roman Catholic and Orthodox Protestant groups placed Christian views on the public agenda. The great theme of socialism, which appeared during the international crisis of the 1880s, was the enfranchisement of the workers. Thus, broadly, four ideological movements had taken shape by the end of the nineteenth century: liberalism, political Catholicism, political Protestantism, and socialism. Each formulated its own analysis of the problems of Dutch society and its own route to the future. Let me briefly review the various ideas about the role of technology in society.

Dutch liberalism had its roots in the tolerant brand of Calvinism that had become prevalent among the ruling class of merchants after the revolt against the Habsburg Empire in the sixteenth century. It was the ideology of the ruling elite between 1850 and 1880, and it remained prominent among entrepreneurs, bankers, and a large part of the bureaucracy well into the twentieth century (Stuurman 1983: 297). In matters of economics and technological development, liberals generally held to the doctrines of the so-called classical economists: McCulloch, Ricardo, Say, Bastiat, and so on (Boschloo 1989: 45–47, 199). Their journals reported enthusiastically on new machines and factories in other countries and urged Dutch entrepreneurs to introduce these innovations in the Netherlands. Machines, it was argued, produced cheaper and better goods and lightened the burden of work. Besides, if Dutch entrepreneurs refused to innovate, the Netherlands would remain a backward country and the Dutch market would be flooded with industrial products made elsewhere.

After 1870, young liberals started to write extensively about the social consequences of industrialization: child labor, excessive working hours, bad hygiene, and so on. They argued that these problems should not be attributed to industrialization itself but to "human ignorance and immorality, obsolete institutions and laws." Social legislation could amend most of these problems, while air pollution and bad working conditions in factories could be eliminated by technical means.[3] The technically oriented *Tijdschrift van de Maatschappij ter bevordering van Nijverheid*, published by an organization comparable to the German *Gewerbevereine* (trade unions), wrote about technical solutions to air pollution and poor factory hygiene. In short, the intellectual appropriation of industrial technology never presented a great problem to the liberals. An important reason for this is probably the late industrialization of the Netherlands. By the second half of the nineteenth century, foreign examples had shown that industry could bring unprecedented welfare and that social legislation could be effective in correcting the worst social effects. Both technical and social advances were seen as signs of the progressive development of human ingenuity and of humanitarian values.

Socialism emerged in the Netherlands in the 1880s. The year 1894 saw the founding of the Sociaal-democratische Arbiderspartij, which had as its main goals an extension of the franchise and social legislation. It cooperated with the so-called *vakbeweging* (modern labor unions), which emphasized strong organization and very selective use of the strike weapon. Both the *vakbeweging* and the SDAP were under continuous

---

3. *Vragen des Tijds*, 1876, I: 24–25, *Tijdschrift der Nederlandsche Maatschappij ter Bevordering van Nijverheid*, 1861: 1–41, 1882: 379, 1896: 223.

attack from syndicalist labor leaders, who favored "direct action" (meaning spontaneous and massive strikes). While parliamentary socialism and "modern" unionism were clearly gaining ground, especially after the dramatic defeat of the workers in the great railway strike in 1903, their hold on the workers was still very weak. In 1914, for example, only about 11 percent of Dutch workers belonged to a union (Altena 1993: 279f.). Syndicalism was still popular, especially among casual workers such as those on the docks (Hueting et al. 1983: 24–32).

Practically all socialist leaders, in the Netherlands as in other countries, were strong advocates of technological innovation, which they considered capitalism's major contribution to human progress (Sieferle 1984, chapter 9). Although workers were the main victims of industrial production, their socialist leaders endlessly repeated the message that these problems were due not to technology but to its uses under capitalism. In the new socialist society, the tremendous apparatus built up under capitalism would be employed for the benefit of all. In socialist publications, from the revolutionary *Recht voor Allen* to the parliamentary *De Nieuwe Tijd*, the tone was one of unwavering optimism and assurance: ". . . our watchword should not be 'away with machinery,' but 'away with the capitalists and capital to the workers.' "[4] One prominent SDAP leader even argued that it was in the best interest of the working classes that strikes against new machinery be lost.[5] Resistance to modern technology was routinely condemned as "reactionary." Instead, it was argued, workers should try to claim their share in the benefits that mechanization could bring, such as shorter work hours and higher wages. It was a view that required a considerable amount of stoicism in the face of suffering brought on by the introduction of new machinery. Before the appearance of socialism in the Netherlands, the rule of liberalism had already been contested by Orthodox Protestants and Roman Catholics. Orthodox Calvinists in the Dutch Reformed Church had opposed the latitudinarian views of the elite from the seventeenth century on. Two secessions, in 1834 and 1886, had created a spate of Orthodox congregations, mainly recruited from small farmers and the lower middle classes in the towns. The most politically articulate of these groups were the Gereformeerden, led by the Amsterdam minister Abraham Kuyper (1837–1920). Kuyper turned religious opposition into a powerful socio-political movement, founding the newspaper *De Standaard* in 1872, the Antirevolutionaire

4. *Recht Voor Allen*, June 25, 1888.
5. *De Nieuwe Tijd*, 1908: 870.

Partij (the first modern political party in the Netherlands) in 1879, and a university in 1880. Kuyper's goal was to re-Christianize society. He believed that all the afflictions of modern society stemmed from the fact that people had turned their backs on the orthodox faith, a process that had been greatly enhanced by the French Revolution (hence the name of his party). He therefore declared war on the modernism that was prevalent in Protestant theology and at the universities and on the godless views of the liberals and socialists.

In his early career, Kuyper saw modern technology as a part of the modern culture that had to be attacked, but by 1900 he had changed his mind. He now claimed that man's dominion over nature was a divine command, stated in Genesis, where Adam was told to "fill the earth and subdue it." Adam had gone about this task without effort in the Garden of Eden, but after the Fall the relationship between man and nature had become hostile. Man had survived only because God had saved his intellectual powers from the destructive consequences of the Fall. This enabled man, slowly and arduously, to regain his original mastery over nature by means of scientific and technological progress. Such progress was not to be seen as the restoration of Paradise. Because of man's sinful nature, increasingly advanced means had been put to evil purposes. But the dark sides of modern life were not to be attributed to technology, which was the product of the human intellect, God's greatest gift to man. Perfecting our mastery over nature remained a divine command. The social problems that accompanied the introduction of industrial machinery were, Kuyper said, ultimately caused by the liberal-capitalist system. Like all other aspects of the modern world, they could be overcome only if society returned to the fold of orthodox religion. This was Kuyper's ultimate mission.

Roman Catholics had been the Netherlands' largest religious minority, comprising about 30 percent of the population, since the revolt against Spain in the sixteenth century. Although the French-supported regime that came to power in 1795 gave them equal rights relative to other religions, centuries of discrimination had left their mark: Catholics were heavily underrepresented at the universities, in Parliament, and in the bureaucracy. After the introduction of the liberal constitution of 1848, they started to build up their church organization and became active in politics, especially in the debate about state subsidies for Catholic elementary schools. They did not have a charismatic leader comparable to Abraham Kuyper, however. Intellectual leadership was entirely in the hands of a large class of priests (one of the consequences of the long period of discrimination), whose thinking about man and society closely followed the latest developments in Rome and especially the social writings of German Roman Catholic thinkers, such as Pesch and Hitze (Gribling 1975).[6]

The development of Roman Catholic ideas about technology and society runs remarkably parallel to Kuyper's. Here too, a conversion took place from a very critical to a very positive stance. This was part of a reorientation of Vatican policy toward the modern world prepared by French and German priests in industrializing areas and confirmed by Leo XIII, who became pope in 1878 (Camp 1969, chapters 1 and 2; McSweeney 1980, chapter 3.). The new social doctrine of the church, which was based on the philosophy of the thirteenth-century Dominican friar Thomas Aquinas, said that Catholics should no longer shun modern culture (as they had been told to do by Leo's predecessor in the *Syllabus Errorum* of 1864) but, rather, should create a Catholic version of it: there should be Catholic scientists, and Catholic workers should have their own trade unions. Like Kuyper's neo-Calvinism, neo-Thomism extolled human ingenuity as God's greatest gift to man and technological progress as a natural—and therefore good—outcome of this. The problems of industrial society were to be blamed not on machines but on the uses to which machines were put under the prevailing liberal-capitalist system. Catholics should therefore both participate in the development of technology *and* work toward a better (corporatist) order.

The "conversion" of Roman Catholic and Orthodox Protestant leaders was due to their realization that by opposing modern technology and industrialization they would marginalize themselves. They therefore adopted elements of liberalism and socialism, hoping to make both superfluous for Christians; and among these elements was technological optimism. The result was a remarkable convergence of opinions about the social meaning of technology toward the end of the nineteenth century.[7] Through various processes of intellectual appropriation, technology was bracketed from discussions about social issues. The groups that dominated public discussion all accepted modern machinery, although they embedded this acceptance in different discursive frameworks.

6. I do not, in this connection, substantiate Gribling's (1975: 24) claim for the originality of Aalberse and Nolens.

7. For a more complete analysis and explanation of this convergence, see van Lente 1992.

Technology came to be considered an arsenal of neutral means with a highly positive potential. This potential would be fully realized only, it was argued, in a well-ordered society, for which left-liberal, socialist, and corporatist (Roman Catholic and Orthodox Protestant) alternatives were proposed. An independent conservative movement did not exist in the Netherlands, although there were conservative tendencies in the Christian parties and in the older generation of liberals. This is significant, because in other countries conservatives were among the most important and consistent critics of industrial society—one thinks of Disraeli, Carlyle, and Riehl (van den Berg 1980, chapter 11; Sieferle 1984, chapter 11).

The Grain Game: Groups and Interests in Rotterdam's Harbor

The conflict over the grain unloaders was a complicated affair involving many different groups with diverse interests. In order to understand the various groups' reactions to the new machinery, we have to acquaint ourselves with their positions in the trade and handling of grain and with the explicit and implicit "rules of the game" (Mol 1980: 106–132; Schilthuis 1918: 127–156; Serton 1919: 11–49, 123f., 154–180; van der Waerden 1911: 222ff.).

The grain trade had been an international affair for centuries. Rotterdam became an important port for grain only after 1870, as a consequence of the rapid rise of the region of Germany known as the Ruhrgebiet, which had the Rhine as its main connection to the sea and Rotterdam as its main seaport. At the same time, grain production in the United States, Argentina, and the Black Sea area increased tremendously, and transportation was facilitated by railways and steamships. In 1872 the access of Rotterdam's harbor to the sea was greatly improved by the opening of a new canal. Rotterdam became the funnel through which grain from all over the world flowed to the almost insatiable Ruhrgebiet. Its position was contested, however, by other ports on the North Sea—especially Bremen, Hamburg, and Antwerp. Because grain was usually transported by irregular services ("tramp shipping") rather than by liner companies with fixed schedules and ports, the routes by which grain reached its customers could be changed easily (van Ijsselstein 1914: 3–10).

The complications of the trade and transshipment of grain were due to mainly two factors. The first was the nature of the product. Grain is a perishable and very valuable bulk good. Therefore, its transportation and transshipment require more than the usual care and are closely supervised by both exporters and receivers. As a consequence of changes in harvests, both the demand for grain in the importing countries and the supply from various countries vary much more than is the case with other bulk goods. As a result, the grain trade was an unusually nervous and uncertain business. The second factor was the long chain of intermediary persons and institutions between producers and consumers. Because of the high cost of overseas transportation, grain was shipped only in large quantities. Exporting firms bought grain from many different producers and sold it to European importing firms, who in turn sold it to flour mills and local wholesalers. Sea transportation was taken care of by a shipping firm, which acted by order of the exporters. There were several shipping firms in Rotterdam, many of which also acted as brokers for colleagues in other countries. Inland transportation from the seaport to the importing firm was the responsibility of the importer. The importer delegated this work to a man called a "factor," who hired lighters at a shipping office.[8] Thus, it was at the seaport (in our case, in Rotterdam) that the grain changed hands from the exporter to the importer. This involved transshipment, taking samples, and weighing.

Before the arrival of the pneumatic grain unloaders, the transshipment of grain was done mainly by hand (although some firms used bucket elevators). The shipping firm entrusted this work to a master stevedore, who hired a gang of dockworkers. These men shoveled the grain into baskets, which were then hoisted onto a ship's deck by means of a winch. On deck the basket was weighed and its contents were transferred to a bag, which was then carried into the board of the ship. There the bag was emptied into a chute which ended above the hold of the lighter.

The weighing of the grain was, of course, very important. The importer had to pay the captain, who represented the exporting firms, for the amount of grain he had ordered. If the amount received, as measured by the weigher, was less than the amount ordered (as was usually the case), the importer could claim a repayment from the exporter. The captain was paid the freight dues for the amount weighed. Before 1866 this weighing had been done by sworn officials employed by the city, who formed a kind of "guild." But this "guild" was dissolved in the general trend toward liberalization of the economy. After that, the

8. Railways were hardly used, since this cost twice as much as transportation by ship (Serton 1919: 34).

importers employed weighers who worked for private weighing firms. This meant, of course, that they were not entirely impartial. In response, exporters created controlling firms to supervise the weighing in their name.

Work on board the grain ships, like most manual labor in the harbor, was hard and unpleasant.[9] The irregularity of the trade put its stamp on the working conditions. Regular jobs hardly existed, and many workers were unemployed for long periods of time. If one had a job, it was likely to mean 24 to 36 hours of continuous labor. Accidents were frequent, but medical care and compensation in times of illness hardly existed. Because laborers were easily recruited from the surrounding countryside, wages and working hours were entirely dictated by the employers. The master stevedores were especially hated by the workers for their high-handed cruelty. It was estimated that there were about 2000 grain workers (including weighers and controllers) in Rotterdam in 1905 (estimates of the total number employed vary between 10,000 and 14,000 in period 1905–1914).[10] Their working hours were usually not as long as those of, say, ore workers, and their wages were a bit higher.[11]

A large seagoing vessel usually contained parcels of grain from several exporters going to several buyers. Each party had its representatives—weighers, factors, and controllers—on board the unloading ship. The unloading procedure usually involved more than a hundred people. It looked, in the words of one grain merchant, like a "seething anthill" (Uyttenbogaard 1928: 284). The haste and confusion of the procedure offered all of the parties many opportunities for embezzlement (Graswinckel and Ott 1973: 30f., 51; Mol 1980: 106–112, 121f.; Schilthuis 1918: 143f.).[12] Some exporters mixed sand with the grain. Others claimed too high a weight for their parcels and then delayed repayment. Importers instructed their weighers to manipulate the weighing, which was not difficult to do on a ship that was rolling on the waves and shaking because of its running engines. Many of the factors took oversize samples, which they subsequently sold for themselves. Dockworkers also took

9. Spiekman 1900, 1907a. See also Algemeene Havenarbeidersvereeniging, "Streven naar verbetering" and "Een noodkreet der Rotterdamsche havenarbeiders," April 3, 1905, Gemeente-archief Rotterdam.

10. *Het Volk*, November 16, 1905; *Katholiek Sociaal Weekblad*, October 26, 1907; Spiekman 1900: 119; Spiekman 1907a: 605; van Ijsselstein 1914: 7.

11. Cf. Spiekman in *Het Volk*, November 16, 1905.

12. See also *De Havenarbeider*, May 19, 1906.

some grain, but usually only a little. (If caught, a dockworker went to jail for a month, leaving his family without an income.)

There were conflicts of interest between exporters and importers and between importers and the shipping companies. Since a ships' owner was paid for the amount of grain delivered, he was hardly happy being dependent on weighers and factors. And because the harbor dues were about half of the transportation costs, ship owners wanted to have their vessels available for new freight as soon as possible (Serton 1919: 20). The grain merchants, on the other hand, were not in such a hurry. It had taken about three weeks for the grain to arrive from the Black Sea or La Plata, so whether the transshipment took another one day or another three did not make much difference to them.[13] Besides, since the grain trade operated on such a fluctuating market, importers sometimes delayed selling their parcels until the last moment. It could therefore be convenient for them to keep their grain on the seagoing vessel for another day or two (Graswinckel and Ott 1973: 50).[14]

The pneumatic grain unloader could solve many problems in one stroke (Serton 1919: 39–44; van der Waerden 1911: 222ff.).[15] The grain was pumped from the hold of the seagoing ship through four tubes into a tank in the tower of the machine, where it was weighed automatically and then released into several tubes leading to lighters. The capacity of the machine was about 150 tons of heavy grain per hour. Clearly, the machine would eliminate a lot of manual labor. Only a few dockworkers were needed to guide the tubes in the hold of the grain ship, while a few men operated the engine and moved the floating unloader about in the harbor. Weighers and controllers were not needed any more. The only work left to the factors was administrative, and this could easily be taken over by either the Unloader Company or the importers. Transshipment thus became much more efficient. Where once it had taken 126 workers seven to eight days to unload a vessel of 6000 tons, two unloader machines could do the same in two days with 14 men each (Serton 1919: 42). This efficiency and the "objectivity" of the weighing procedure must have been especially appealing to the ship owners and the brokers.

And the new machines offered many other improvements in the

13. *Nieuwe Rotterdamsche Courant*, May 11, 1907.

14. See also Gemeente-archief Rotterdam, Archief van de Kamer van Koophandel no. 165, "Advies van de Commissie."

15. For technical details see Gemeente-archief Rotterdam, Archief Graan Elevator Maatschappij, no. 11.

transshipment of grain (Cocheret 1933: 12ff.; Voogd 1907: 5f.).[16] Since the tubes required only a small opening of the hatches of the grain ship, the grain was hardly exposed to the open air, so transshipment could go on regardless of wind and rain. Whereas much grain was lost when poured into an open chute, this could not happen with the unloader.[17]

Enter the Machine[18]

The first initiative to introduce pneumatic grain unloaders in Rotterdam came not from ship owners and brokers, as one might have expected, but from a group of German importers, the Verein Deutscher Handelsmüller, who sent a delegation to Rotterdam in 1901. Pneumatic grain unloaders had been in use for some time in London and had recently been introduced in Hamburg, Bremen, Genoa, and New York.[19] Rotterdam, said the German merchants, should not lag behind. In an address to the Rotterdam Chamber of Commerce, they spelled out the advantages of the machine. They pointed especially to faster transshipment, the "objective" weighing procedure, and the possibility of cleaning the grain and thus avoiding the payment of import duties for useless weight.[20] The enthusiasm of the importers had to do with the trouble they were having with Russian and Romanian exporters, whom they accused of mixing a lot of dirt with the grain and having their controllers manipulate the weighing. Four years later, in 1905, the German and the Dutch importers decided to create a charter that would be imposed on the exporters on the Black Sea to rule out these irregularities (Everwijn 1912: 643–646; Schilthuis 1918: 143).

The Chamber of Commerce was not impressed by the Germans'

16. See also *Nieuwe Rotterdamsche Courant*, May 22, 1907.

17. A possible advantage was that the machine cleaned the grain automatically by removing dust and other light admixtures. This was, of course, not in the interest of the exporters, who would have to pay higher restitutions. Therefore the unloader had a device to mix the dust again with the grain before weighing (Serton 1919: 41).

18. For good accounts from the viewpoint of the employers, see Cocheret 1933, Uyttenbogaard 1928, Voogd 1907; see also two anonymous articles in the newspaper *De Nieuwe Nederlander* that also appeared in *De Maasbode* (November 12 and 14, 1905). For accounts from the viewpoint of the workers see Berg 1906a, Mol 1920, Mol 1980, and Brautigam 1956.

19. Reports in Gemeente-archief Rotterdam, Archief Graan Elevator Maatschappij, no. 1.

20. See also Gemeente-archief Rotterdam, Archief Nederlandsche Veem, notulen bestuursvergadering 8 augustus 1901. I owe this reference to Hugo van Driel.

arguments, but it recommended the introduction of the machine on the grounds that automatic weighing and lower labor costs would attract commerce to Rotterdam.[21] The records show no great concern on the part of the Rotterdam entrepreneurs about lagging behind other ports with regard to unloading equipment. This is understandable, since even in London, Hamburg, and New York most of the work was still done by hand. New York, for example, though far behind Hamburg in port equipment, had a much better reputation for quick dispatch, because its dockmen worked longer hours and at higher speed (Lovell 1969: 28). The Chamber of Commerce recommended that the city of Rotterdam introduce grain unloaders as it had done with electric cranes, but apparently nothing came of this.[22] Then the Nederlandsche Veem, a storage and shipping firm, took the initiative (van Driel 1993: 21ff.; van Driel 1992: 98ff.). It turned out to be hard to raise capital for a grain unloader company. Neither the German nor the Dutch importers would risk any money on it—and, as I have mentioned, quick dispatch and objective weighing were not really in their interest. When the Grain Unloader Company was finally founded, in March 1904, it was financed largely by Rotterdam ship owners and brokers (although three factors and one stevedore company also took some shares), whose most important motive was quick dispatch.[23] The capital raised was sufficient for only two machines (Cocheret 1933: 20, 50).[24] It was decided to buy floating unloaders, which could be moved all around a ship in the new Maashaven (the harbor in Rotterdam, built in the late 1890s).

21. Gemeente-archief Rotterdam, Archief Kamer van Koophandel, no. 113, letter no. 165, advice of the committee of the Chamber of Commerce about the importance of grain unloaders for the Rotterdam harbor.

22. In the minutes of the municipal council and in the archives of both the municipal council and the city administration the whole conflict is hardly mentioned at all, which shows that local politicians regarded this as an affair between employers and workers. See Gemeente-archief Rotterdam, "Handelingen van de Gemeenteraad," 1900–1907; Archief Gemeenteraad en college van Burgemeester en Wethouders, secretarieafdeling Algemene Zaken en kabinet van de burgemeester, no. 5698, Index op uitgaande stukken van de burgemeester; 1141, Index op notulen B. en W. 1902–1922; 674, Notulen van geheime raadsvergaderingen 1881–1907.

23. Scheepvaart Vereeniging Zuid, July 22, 1905.

24. That a lack of capital was the main reason for buying only two machines is not explicitly stated in the sources, but it may be inferred from the arguments in the letter of February 1, 1904, in the archives of the Graan Elevator Maatschappij at the Gemeente-archief Rotterdam (no. 2A) and from the arguments put forward later by the company.

Both the Chamber of Commerce and the new Grain Unloader Company anticipated trouble with workers, master stevedores, and factors, all of whom were threatened by the machines. The company decided to cooperate as much as possible with the master stevedores and factors, in order to prevent them from siding with the workers. They offered to put the machines at the disposal of the stevedores and factors and to give them a share in the profits. In other words, they wanted to integrate the machine into the existing system of transshipment. Experience must, they thought, have made it clear to the dockworkers that resistance to new machinery was completely ineffective. In the long run, the new machinery would attract more freight to the harbor, which would create sufficient employment.[25]

The Rotterdam dockworkers indeed had some experience with collective action (Spiekman 1900: 117–140; Smits 1902: 100f., 152–155, 195–200; Mol 1920; Van 't Wel 1986: 66–85). In 1882, a new bucket conveyor for grain had been destroyed by fire after a strike (Crol 1947: 24; Graswinckel and Ott 1973: 24, 37).[26] In 1889, inspired by a great strike in the harbor of London, Rotterdam's dockworkers held their first massive strike, and got a raise in wages. In 1896 the introduction of electric cranes had provoked a strike by the oreworkers, which had failed. In 1900 a great strike for shorter working hours and better working conditions had also failed. After the short and unsuccessful actions that had followed the introduction of new transport machinery for the transshipment of coal in 1903 and 1904, many jobs had been lost.

The pattern of these actions was always the same, and very similar to that in other great European ports around 1900. They came suddenly, were hardly planned, exhibited a remarkable solidarity between different categories of workers, and never lasted longer than two weeks (most lasting a much shorter time). Labor unions hardly existed. The prominence of casual labor (which induced a sense of individualism), the constant competition for jobs, and the poverty of the men all stood in the way of permanent organizations with ample strike funds. Therefore, when the workers took action, they tried to overwhelm the employers by a massive and sudden attack—what syndicalists called "direct action."

25. Gemeente-archief Rotterdam, Archief Graan Elevator Maatschappij, no. 2A, letter of February 1, 1904.

26. This "sad experience" was referred to by the Chamber of Commerce in its advice. Whether this was a case of arson, and if so by whom, is not clear from the sources. See Gemeente-archief Rotterdam, Politie-archief, no. 138, 1883.

The success of some of these actions, such as the one in 1889, confirmed their faith in this tactic. During a strike, organizations would be set up and would be joined by hundreds or even thousands of workers, only to shrink afterward and often to disappear. If any concessions were granted by the employers (as happened in 1889), these were silently withdrawn in succeeding years. During a strike, the employers could easily break the resistance of the workers by recruiting men from the villages in the region or from other ports, such as Hamburg and London (Mol 1920: 527, 530, 538; Jansen 1979: 54ff., 80–87; Broeze 1991: 169, 176, 184, 194; Lovell 1969: 97; Grüttner 1984: 245–251).

Building the grain unloaders took more than a year. In July 1905 they were finished, and in August the first grain ship was unloaded by machine. Unfortunately for the Unloader Company, the weighing apparatus did not work. Because no provision had been made to break the fall of the grain into the scale, its indications were far too high.[27] Since this could not be repaired quickly enough, the company had to install a container on deck of the ship from which large batches of grain could be weighed by hand in a weighing apparatus. The installation of this apparatus took six weeks.

The spectacle of the iron monsters at work had clearly shocked the workers. During the six weeks in which the machines were idle, weighers, controllers, factors, and master stevedores—the groups whose work the machine threatened most to eliminate—all created their own organizations. At a meeting held on August 13, A. C. Wessels, the Social Democrat leader of the largest dockworkers' union, the Algemeene Havenarbeidersvereeniging, tried in vain to unify these organizations and to restrain their demands. Some favored "direct action," others a more limited strike. The meeting ended in turmoil when a well-known coalworker and agitator named Ooykaas wildly accused several leaders of vagueness and avoiding the real issue.[28] The new organizations demanded that no

27. This was ascribed to the hasty installation of the machines, whose delivery had been delayed. It is somewhat surprising, since the German firm Luther had delivered machines with good weighing apparatus before. The Graan Elevator Maatschappij used hand weighing apparatus at least until the end of the conflict in 1907. See printed letter to German importers, dated March 23, 1907, Gemeente-archief Rotterdam, Archief Graan Elevator Maatschappij, no. 55.

28. *De Maasbode*, August 16, 1905; *Rotterdamsch Nieuwsblad*, August 15, 1905; *Nieuwe Rotterdamsche Courant*, August 14, 1905. Ooykaas's boss, Van Beuningen, also describes him as a hot-tempered man who called himself a Christian anarchist (which places him close to Sam van den Berg). See Gemeente-archief Rotterdam, handschrift no. 326, p. 28f.

workers be fired and the unloaders be used only to supplement unloading by hand. The company offered a 10 percent raise in wages in exchange for acceptance of the machines and the introduction of two new ones. It could, of course, not guarantee that the same number of workers would find work in the transshipment of grain. These offers were not acceptable to the workers' organizations, but when all was said and done only the weighers, the best-organized group, were prepared for action. They called a strike on November 3, and the next day not one of the 450 weighers appeared on a ship. The controllers immediately joined the strike. The personnel of the Grain Unloader Company continued to work, and several ships were unloaded by hand during subsequent weeks. Some captains had the grain transshipped unweighed, because they could not afford to leave their vessels idle for an unknown length of time. The costs of the lighters and the weighing that had to be done later would probably have to be paid for by the importers, who were expected to challenge this procedure in court.[29] It is impossible to tell how many workers besides weighers and controllers struck and what part of the grain trade was stopped; however, it is clear that the weighers, by their concerted action, had created considerable stagnation and had caused a panic among the importers, especially those in Germany.

When the German importers heard about the strike, they immediately sent a committee to Rotterdam's mayor, who refused to interfere in the conflict. Then, after negotiation, the importers reached an agreement with the factors and the weighers. During the next half-year they would not accept any grain that had been weighed by machine. Thus, the weighers had won a fast victory. On November 21 they were at work again. The Unloader Company was forced to lay up its machines.

The action of the German importers is not difficult to understand. To them, the strike came at a very unfortunate moment: just before the onset of winter, when the Rhine was often frozen, making transportation impossible. On the first of March, the German tariff on grain would be raised. Therefore, they had no time to lose. Since the two unloaders could transship only 10 percent of all the incoming grain at most, the importers would be dependent on the weighers for some time to come, and had no choice but to negotiate with them (Voogd 1907: 11). They also complained that the unloaders weighed inaccurately, but this was probably a spurious argument. In Rotterdam, said one observer, "the sparrows [were] squeaking from the rooftops" that the unloaders

29. See reports in *De Maasbode* during the strike about ships being unloaded.

weighed only too accurately for the importers (Cocheret 1933: 60). The weighers, who worked for the importers, used to write up far lower weights than the importers were receiving (up to 20 percent, according to some), so the importers had a vested interest in the continuation of the old procedure of transshipment (Cocheret 1933: 60, 65, 72, 119; Voogd 1907: 27; Mol 1980: 107f.).

Though understandable, the aforementioned agreement was probably illegal. Most of the weighers were employees of weighing firms; the most prestigious of them were with the Comité van Graanhandelaren te Rotterdam (Committee of Dutch Grain Merchants). The German importers had completely bypassed these firms and their Dutch colleagues. This explains why most Dutch grain merchants refused to sign the contract. They were rather indifferent to the unloader, but they could not accept that "their" weighers were, together with the Germans, imposing the order of the grain trade (Cocheret 1933: 48).[30] Another legal problem was whether the grain importers had the right to determine the method of delivery. Their main argument—that they owned the grain—was not very strong. It was the captain (and hence the ship's owner) who, on the basis of the bill of lading, determined the manner of transshipment. Many of these charters included a clause that transshipment should proceed "as fast as steamer can deliver." This could be taken to imply delivery by machine, but whether this had to include weighing was a moot question; it was only decided in court years later, when it was no longer in dispute.

In the meantime, the Unloader Company was at a loss about what to do. It realized that it would have been better to introduce twelve unloaders instead of only two, so that the resistance of weighers, workers and factors could have been swept away in one stroke (Graswinckel and Ott 1973: 48). It considered buying more machines,[31] but it did not do so, probably because obtaining the necessary capital would have been at the time. It also considered selling the machines. (In fact, they were offered to Hamburg.) Finally, in April 1906, a master stevedore named Thomsen hired the unloaders for half a year, offering to keep as many men employed as if transshipment were to be done by hand. Because

---

30. In fact the position of the Dutch importers was very awkward. They had, after long negotiations, reached an agreement with their German colleagues about the Black Sea charters, which they did not want to jeopardize. The unloaders were not really in their interest (Graswinckel and Ott 1973: 49f.).

31. See Gemeente-archief Rotterdam, Archief Graan Elevator Maatschappij archive, no. 3, for several quotes dated February and March 1906.

the machines would do most of the work, the workers would receive half of their present wages, and during the next three years no new unloaders would be introduced.[32] The Algemeene Havenarbeidersvereeniging (General Dockworkers Association)—the Rotterdam section of the socialist-oriented Nederlandsche Scheeps- en Bootwerkersbond—urged the workers to accept this offer on some conditions,[33] but the assembled unions turned it down. They said the dockers did not want to work less if that cost them half their wages. They were afraid Thomsen was trying to establish a monopoly in grain transshipment, and that in the end he would introduce more machines and fire the workers.[34] When Thomsen was unable to find any men to work for him, the contract was not renewed.

The Debate

The first working men's association to react to the coming of the new machines was the Algemeene Havenarbeidersvereeniging (AHAV). After the lost strike in 1900 it had shrunk to only a few hundred members, but during the conflict over the grain unloaders it grew again. In 1907 between 1600 and 2000 dockworkers belonged to it.[35] At the beginning of April 1904 its weekly paper *De Havenarbeider* (The Dockworker) reviewed the innovations that had recently been introduced in the harbor, all of which had eliminated labor. Reminding the workers of the fruitless attempts of the English machine breakers to eliminate textile machinery almost a century earlier, *De Havenarbeider,* calling the mechanization of work part of the inevitable progress of society, urged the workers not to try to destroy the machines. Eventually, in a socialist society, this machinery would be employed for the benefit of all, *De Havenarbeider* suggested. Before such a society was realized, the workers should unite

32. Gemeente-archief Rotterdam, Archief Kamer van Koopkandel, no. 118, printed letter April 3, 1906 (no. 100).

33. *De Havenarbeider,* April 14 and 21, 1906.

34. *Toenadering,* 1/4, May 17, 1906; *De Volksbanier,* April 26, 1906; *De Havenarbeider,* 14, April 21, 1906. One must remember that master stevedores and dockworkers were often very hostile to one another. The workers distrusted these employers more than any other group (cf. Jansen 1979: 31–37).

35. The estimate of 1600 is from the *Katholiek Sociaal Weekblad* of October 26, 1907. Spiekman claimed 2000 "regularly paying members" in *De Nieuwe Tijd* (1907, p. 740).

to claim their share in the advantages of mechanization—that is, shorter working hours and higher wages. The paper warned that as long as the Rotterdam laborers remained indifferent toward the union they would lose every battle with the employers.

Thus, the problem of the grain unloaders was discussed in terms of the common socialist ideology of technology,[36] and the AHAV was supported in its opinion not only by SDAP and labor union leaders such as F. M. Wibaut,[37] L. M. Hermans, N. van Hinte, and J. Oudegeest[38] but also by more radical socialists such as H. Koltek,[39] F. Domela Nieuwen-huis,[40] and H. Gorter.[41] This official standpoint was not shared by all members of the AHAV. At a meeting on August 30, a majority voted in favor of a strike against the unloaders (van den Berg 1906a: 13; Mol 1980: 176). Nevertheless, the AHAV condemned the strike when it broke out two months later, although it decided to support the workers finan-cially out of solidarity. In the meantime it constantly urged them to join the organization.

The articles in *De Havenarbeider* suggest that the AHAV's leaders endorsed the official socialist doctrine but felt that they had to support the strike if they did not want to lose the respect of the workers. An important factor in the leaders' taking this position was probably that the strikers found an unexpected ally in the person of Hendrik Spiekman,

36. The German publication *Hafenarbeite* (July 20, 1907) called this battle against a machine that had already been established in Hamburg for a few years *eigenartig* (peculiar) and not *volkswirtschaftlich richtig* (economically pertinent), but hoped that the strikers would be successful, because this would strengthen their organi-zation to such an extent that in the future they would not have to strike against new machinery (but could claim its advantages, *Hafenarbeite* implied). See Gemeente-archief Rotterdam, Archief Graan Elevator Maatschappij, no. 9A. Grüttner (1984: 47) cites the German labor union leader Döring, who used more forceful language to condemn the Rotterdam strike, calling the struggle against the machine—this "Errungenschaft der Kultur und der Technik"—"etwas Unverständliches, eine Barbarei."

37. See Wibaut 1905.

38. See Cocheret 1933: 44.

39. *De Havenarbeider*, October 14, 1905.

40. Berg (1906a: 9) and Mol (1920: 545) both report of a meeting of the Socialist propaganda club at which Domela spoke. He is cited similarly by the entrepreneur Van Beuningen, who was also present at this meeting, in his manuscript autobiogra-phy (Gemeente-archief Rotterdam, manuscript no. 326, p. 27).

41. Cited in *Voorwaarts: Weekblad voor de Arbeiderspartij in Zuid-Holland*, November 4, 1905.

the most popular SDAP leader in Rotterdam. In response to a survey held a few weeks before the strike by AHAV's chairman, A. C. Wessels, Spiekman had said that if the workers had a real possibility to block the introduction of a machine that would seriously harm their interests they should do so. In a series of articles in the SDAP newspaper *Het Volk* he defended this position, which, he realized, would be regarded as "reactionary" by most socialists.[42] Often, Spiekman said, the introduction of machinery cannot be avoided, because of foreign competition. It can also, if it lightens the burden of labor, be an improvement for the workers. If mechanization proceeds gradually, the loss of jobs will be compensated for by the fact that new firms, new products, and new markets create more work. But in the case of the grain unloaders none of these arguments were valid. Because of their rapid introduction, workers who had become superfluous would not easily find new jobs. The competition argument did not hold, according to Spiekman, because during the last few years business at Rotterdam's harbor had, without machines, grown much faster than business at competing ports. With its large supply of labor, Rotterdam could easily handle a further increase in transshipments. Therefore, the Unloader Company's only motive could be to minimize the income of workers, master stevedores, factors, and weighers, whose work would be made superfluous by the machine. This was unacceptable. If, however, the machines were to be introduced only in order to speed up transshipment, while the same number of workers were employed, Spiekman had no objections to their introduction.

Spiekman was immediately criticized by his comrades for this departure from socialist doctrine, and he found no support at all from other SDAP leaders. Wessel replied in *Het Volk* that mechanization made transshipment so much cheaper that all harbors would eventually be forced to introduce machinery. The argument that in recent years Rotterdam had outstripped other seaports that used unloaders did not impress Wessel: when more machines are introduced, he said, the effects will soon make themselves felt.[43]

Why did Spiekman take this unusual position? Like other "modern" union leaders, he had pleaded time and again for organization (he himself was a compositor, one of the best organized trades). He was exasperated by the fact that unions built up during strikes disintegrated

42. *Het Volk*, November 16, 18, 21, 1905.
43. Ibid., November 28, 1905.

shortly afterward (Spiekman 1900: 117, 124, 138). He therefore felt an urgent need to support and consolidate the organizations that had emerged during this conflict. This could be done only by bringing them together in a union such as the AHAV. But since Wessels and other Social Democrats could not convince the workers with vague slogans like "here with the machine," Spiekman attempted, unsuccessfully, to reconcile the aims of the strikers with socialist doctrine.[44]

The Orthodox Calvinist dockworkers' union Toenadering was founded in 1900 by members of the Nederlandsche Scheeps- en Bootwerkersbond who refused to attend meetings on Sunday.[45] It was closely connected with Patrimonium, the national organization of Orthodox Protestant workers. Its opposition to the great railway strike in 1903 had earned Toenadering a bad reputation with the socialists.[46] At its foundation it claimed to have about 1000 members,[47] but by 1905 this number had probably dwindled to a few hundred.[48] The name of this union, which means "rapprochement," proclaimed its aim. Rejecting the principle of class conflict, it aimed to negotiate with employers. Early in 1905, it sent a delegation to London to inquire about the new machinery. Immediately after the return of the delegation, Toenadering opened negotiations with the employers. In August 1905 it proposed that the unloaders be used only to transship loads over 500 last[49] and

44. Spiekman's position illustrates nicely the conflict between militancy and pragmatism that, according to Broeze (1991: 178), was typical of maritime labor organizations all over the world in this period.

45. See the memorial volume *Gedenkboekje uitgegeven ter herinnering aan het 25-jarig bestaan van "Toenadering." afdeeling Rotterdam van den Nederlandschen Bond van Christelijke Fabrieks- en Transportarbeiders 10 juli 1900—10 juli 1925.* s. l. s. a. (1925) and the weekly paper *Toenadering*, which began to appear in February 1906; both are at the International Institute for Social History in Amsterdam.

46. See e.g. *De Havenarbeider*, January 14, 1905.

47. Gemeente-archief Rotterdam, collection "pol. vakbonden alf.," bulletin 1900. The relatively large size of this union can be explained by the fact that many Rotterdam workers were recruited from the villages to the south and south east of Rotterdam, which is part of the Dutch "Bible Belt."

48. According to Mol (1920: 542), Toenadering, with only 70 members after the great strike of 1900, was still the largest union in the harbor. The *Katholiek Sociaal Weekblad* of October 26, 1907, estimated its membership in 1907 at 200; in 1908, according to Teychiné Stakenburg (1957: 19), it was 150.

49. One last is about 2000 kilograms of grain. See Schilthuis 1918: 18, 129.

that wages be increased.[50] Although Wessels of the AHAV found this a good proposal, the dockworkers' organizations turned it down. The employers simply ignored it.

Like the AHAV, Toenadering approached the problem of the unloaders from a general ideology about technology and society. This ideology was clearly and explicitly derived from Abraham Kuyper.[51] Applied to the present case, it implied that the harbor was a working community in which each participant had inviolable rights. Therefore, the introduction of machinery that would have a great impact on the workers should be the subject of negotiations with all persons involved. But the present situation, said the union publication *Toenadering*, was one in which "might is right." The unloaders served exclusively the needs of the employers. The socialists only wanted to reverse the present power relationship between workers and employers, which would obviously lead to new forms of repression. The only true solution was to create order, not on the basis of power, but on the basis of the divine law that God, by His grace, had inscribed in every human heart. For Toenadering this meant the acceptance of the grain unloaders, which, like every new technology, were to be considered gifts from God. It also meant acceptance of the authority of the employers, who had received this responsibility from God. But it also meant that the benefits of the new technology should accrue to the whole harbor community, not just to the employers. A worker could not simply be dismissed from his job, for God had granted him the right to "eat his bread by the sweat of his brow." Since the employers had not responded to the reasonable proposals that Toenadering had made in August, it considered the strike of the weighers justified. Pleading for negotiations between workers and employers, it opposed a general strike "for the time being."[52]

This position was much more radical than that of Patrimonium, which stressed the inevitability of technological progress and supported Toenadering's proposals for negotiations but ignored Toenadering's endorsement of the strike. It condemned the idea of a general strike as "anarchist," while Toenadering had only dismissed it "for the time being." When the newspaper *Patrimonium* suggested that the unloaders

50. *Toenadering*, May 17, 1906. It is hard to know what percentage of the parcels delivered in Rotterdam was over 10 tons and therefore apt for unloading by machine, according to *Toenadering*.

51. See especially *Toenadering*, May 16 and November 28, 1907.

52. *Patrimonium*, November 16, 1905.

would probably not cause much unemployment, since trade in the harbor was increasing, it drew a critical reply from a Rotterdam member, who wrote that machinery would certainly replace much labor in the future and that the employers, by not taking the interests of the workers into account, deserved a strong reaction.[53] Abraham Kuyper ignored the strike in his daily editorials in *De Standaard*, but 10 years earlier he had written—in connection with the strike against the introduction of electric cranes—that such actions were useless.[54] The men of Toenadering therefore found very little support from national Orthodox Protestant leaders.

The Roman Catholic dockworkers also had a small union, called "Kardinaal Manning," with at most 200 members in 1907.[55] It was the dockworkers' section of the Rotterdam department of the Roomsch Katholieke Volksbond (Roman Catholic People's League). At first "Kardinaal Manning" had not officially backed up the strike, but its men had helped the strikers, for example by distributing their manifestoes (van den Berg 1906a: 23). Later this union, like Toenadering, came to support the strike. The chairman of the department, F. J. B. van Rijswijk, explained this position in a long article in November 1905, after the strike had ended. His arguments, often repeated during the following months, were far removed from the official technological optimism of Roman Catholic leaders. Indeed, van Rijswijk's point of view was close to Spiekman's, in that he questioned the necessity of introducing the new machines. Rotterdam's harbor, he argued, was internationally known for its low prices and quick dispatch, and there had never been a lack of workers. The introduction of the unloaders was therefore not in the interest of the harbor as a whole; it was only in the interest of the employers. Their argument that the new equipment will attract more work was spurious, because other harbors, like Antwerp, would be forced to follow suit, thereby increasing competition. Although he did not say so explicitly, Van Rijswijk does not seem to have rejected the unloader

53. Ibid., October 12 and 19, 1905.

54. *De Standaard*, March 25, 1896.

55. "Kardinaal Manning" did not have its own newspaper. Its leaders published in *De Volksbanier*, which for our period is only preserved at the International Institute of Social History in Amsterdam up to June 1907. It is my main source for the following statements. The number of members here mentioned is taken from the *Katholiek Sociaal Weekblad* of October 26, 1907. The same article says that 60 members appeared at a meeting in June 1906; a priest claimed in *De Volksbanier* (June 7, 1906) that the organization should have at least 1000 members.

as such; he wrote that if the unloaders had been introduced 20 years before, when the grain trade started to grow, then they could have been easily integrated into the transshipment system. But as it happened, the growth of the grain trade had led to a tremendous increase in the number of workers, who were now dependent on the transshipment of grain by hand. In view of this, the transition to transshipment by machine should be carried out in close cooperation with workers' representatives, in order to prevent as much as possible the damage that would be done to the workers. That this had not been done showed that the employers were only interested in their own profits. Therefore, wrote Van Rijswijk in conclusion, "may the success of this strike be followed by a complete victory of the workers, in order that the Rotterdam harbor, with its great reputation for speed and cheapness, may not be mechanized for the profit of a few, at the expense of thousands."[56]

The Rotterdam newspaper *De Maasbode*, a mouthpiece of conservative Roman Catholic opinion, repeated the cliché that technological progress "could not and should not" be stopped and held that negotiations should end the problem. In its detailed accounts of the strike, *De Maasbode* simply ignored Van Rijswijk's point of view. So did the other leading Roman Catholic newspaper, *De Tijd*.

Some individuals tried to convince the workers of the need to strike against the unloaders. As far as the sources show, physical violence against the machines was never advocated and was often condemned as useless.[57] An interesting figure whose views deserve some attention was Sam van den Berg, "ethical anarchist," opponent of socialist materialism, and hawker of *Vrede* (*Peace*), a journal of Tolstoy adepts.[58] Van den Berg,

56. *De Volksbanier*, November 23, 1905. A similar position had already been taken by the Roman Catholic candidate for Parliament in one of the Rotterdam districts in September 1905, before the outbreak of the strike (Berg 1906a: 13). Afterward, Jacob Van Term had been denounced by the leading Roman Catholic newspaper *De Maasbode* for his criticism of clerical dominance (apparently unrelated to the *elevator-kwestie*), his political career came to a quick halt and he returned to journalism. See *De Maasbode*, September 10, 1905, and *De Tijd*, September 12, 15, and 16, 1905, and July 12, 1907. The anti-unloader standpoint was repeated in *De Volksbanier* on April 26, May 24, June 7, and June 28, 1906. *De Volksbanier* is not preserved for the period July 1906–July 1908.

57. Their opinions are recorded in *Nieuwe Rotterdamsche Courant* (August 14 and 18, 1905) and in *Rotterdamsch Nieuwsblad* (August 15, 1905). See the anarchist Sam van den Berg's angry reaction to the suggestion that he would have the unloaders towed to the sea in *De Havenarbeider*, June 17 and 24, 1905.

58. See Berg 1906a. See also the defense of this brochure by van den Berg himself in *De Havenarbeider*, May 12, 19, 26, 1906.

who often spoke at public assemblies, was secretary of the Resistance Committee against the grain unloaders (a federation of the newly emerged organizations), and in December 1907 he became the salaried secretary of the National Federation of Transport Workers.[59] At the beginning of 1906 he published a pamphlet about the conflict. Basically, his attitude was similar to that of the Roman Catholics and the Orthodox Protestants.[60] He too pleaded for a just order in which the workers would be respected as much as anyone else. He said that, whereas the basic idea of socialism was that of brotherhood, mainstream socialism had become materialistic and preoccupied with power. Van den Berg portrayed the harbor as a community and the Unloader Company as an intruder that numbered among its shareholders two bankers, a notary, and a wine merchant—people who had nothing to do with the harbor and who were only in this business for the money. (The same point was made in the protest song quoted at the beginning of his article and in the manifesto the Resistance Committee published in the October 14, 1905, issue of *Voorwaarts*. See van den Berg 1906a: 25f.)

The most interesting aspect of van den Berg's standpoint was his criticism of the generally accepted doctrine (shared by socialists, Christians, and liberals) of the inevitable and ultimately beneficial progress of technology. By promoting this doctrine, van den Berg said, union leaders confused and paralyzed the workers, inducing them to accept innovations fatalistically. Technology often did benefit society as a whole, but not necessarily and always. In the case of the grain unloader, it was clear that only the employers would reap the benefits, and therefore the workers should resist its introduction. Since the machine would destroy employment on a massive scale, it made no sense to claim higher wages and shorter working hours: "Wages and hours for whom?" van den Berg asked. It was therefore only natural that the workers protested and disregarded the artificial doctrines of their leaders. The only solution would be the creation of cooperative firms that would fight the capitalistic

59. International Institute for Social History, Archief Nederlandsche Federatie Transportarbeiders, 1903–1908, letter of J. Brautigam to S. van den Berg, December 8, 1907.

60. There was a mutual sympathy between van den Berg and Christian union leaders. *Toenadering* sometimes quoted him admiringly (e.g. on November 28, 1907), and Berg (1906a: 13, 25) quoted the Catholics Van Rijswijk and Term with approval.

ones. When in the possession of the workers, machines would be beneficial to the community.[61]

Social democrats such as Spiekman and Wessels usually lumped Sam van den Berg together with noisy workmen's leaders like Ooykaas and Croiset, who regularly disrupted meetings by loudly accusing moderate union men of being "accomplices of capitalism" and "beasts of prey" and by advocating "direct action."[62] The Social Democrat papers *Voorwaarts* and *Havenarbeider* called them "anarchists," suggesting that they followed the dubious French idea of spontaneous general strikes or even violence to machinery and that they would lead the workers to ruin. Organization, Spiekman repeated, should precede any action.[63]

Second Introduction, Second Strike

Early in 1907 the Grain Unloader Company chose to take the offensive again. There were several reasons for this. Some German importers had indicated that they would accept grain transshipped by machine if it could be guaranteed that no strikes would occur. They suggested creating a new corps of weighers. A British shipping firm had informed its agent, who was on the board of the Unloader Company, that it wanted to use the company's services. And from Hamburg came the news that another five pneumatic grain unloaders had been installed (Cocheret 1933: 67f.). The Unloader Company then made an agreement with nine Dutch importers who were prepared to accept grain transshipped and weighed by machine. The company offered them very low prices and guaranteed to pay for losses they might incur in the case of a boycott (Voogd 1907: 25f.). Next, it created its own corps of weighers and dockworkers, who received relatively high wages. For this they needed the cooperation of the master stevedores, who claimed and received high compensation (Cocheret 1933: 73f.). The cooperation of other employers was therefore

---

61. See Berg 1906a and 1906b and articles in *Vrede* 8 (1905) (p. 118f. and p. 186ff.). The last of these is credited to a "Jan Boezeroen," which may well be a pseudonym of van den Berg. The idea of cooperative firms was also put forward in a short anonymous article in *Sociaal Weekblad* (November 11, 1905).

62. *Rotterdamsch Nieuwsblad*, August 15, 1905; *Voorwaarts*, November 25, 1905.

63. See, e.g., Spiekman in debate with Sam van den Berg about "direct action" against the unloader, *De Bondsbanier*, June 10, 17, and 24 and July 22, 1905. See also *Voorwaarts*, October 14 and November 4, 1905. See also the discussion between the Social Democrat Noordijk and van den Berg about the latter's pamphlet in *De Havenarbeider*, April 7, 1905, and May 12, 19 and 26, 1906.

dearly bought, but it was the beginning of a united front of employers, that would emerge victorious from the conflict. Finally, the company decided to use the unloaders in all cases where the bill of lading included the clause "as fast as steamer can deliver." Importers who would not accept their grain in this way would find themselves required to bear the cost of having their parcels pumped into a "captain's lighter." The German importers were informed of these decisions in a forceful letter, the unwritten message of which was that it was that the owners of the Unloader Company were the bosses in the harbor.[64] The Germans, in reply, demanded the creation of a fund of half a million marks, to be deposited at a German bank, to pay for any losses they might incur. When the Unloader Company chose not to answer, the Germans renewed their contract with the weighers, this time for three years. The contract gave the weighers higher wages but prohibited them to strike.

The new offensive of the Unloader Company was accompanied by long articles in the leading newspapers in which the employers explained their position. Like the other groups participating in the debate, they couched their arguments about the unloader in general ideological terms about technological development. The employers' point of view was entirely in line with the ideas of liberals and leading economists, as discussed above. One of them wrote:

One notices with surprise, that there are still many workers, and not only they, who even in the twentieth century believe that they have to take up the battle of the handicrafts against the progress of technology. But the history of the last hundred years teaches us that a machine that performs a job better, quicker and cheaper than a handicraft inexorably replaces that craft. On the other hand, the growth of commerce creates new jobs. Therefore we have to teach the worker to adapt to technological change.[65]

Another said that the basic question was whether workers should be allowed to block technological progress in the harbor. Without innovation Rotterdam would soon lose its competitive edge to other harbors, such as Antwerp and Emden, and even more workers would lose their jobs.[66] The weighers' organization issued a protest against the creation of a new corps of weighers and dockworkers by the Unloader Company.

64. Gemeente-archief Rotterdam, Archief Graan Elevator Maatschappij, no. 55, letter of March 23, 1907.

65. *Nieuwe Rotterdamsche Courant*, May 12, 1907.

66. Schilthuis, in *Nieuwe Rotterdamsche Courant*, May 22, 1907; cf. *Rotterdamsch Weekblad*, May 18, 1907; Plate, cited in *De Havenarbeider*, July 16, 1904.

The manifesto was signed by the AHAV, Toenadering, and "Kardinaal Manning," among others.[67] When, at a meeting on April 29, 1907, the weighers called a boycott against those importers who had signed the contract with the Unloader Company, they again received the support of all the dockworkers' organizations, including the Protestant and Roman Catholic ones. At the end of the meeting, the *Elevator lied* (the song quoted in the epigraph above) was sung again.[68] A successful strike of grain and ore workers at the port of Antwerp and a dockworkers' strike at Hamburg, both taking place at this time, seem to have been important stimuli (Brautigam 1956: 87).

Once again, the Christian dockworkers' organizations received no support from their more powerful co-religionists. *Patrimonium*, the Orthodox Calvinist workingmen's newspaper, quoted without comment articles from another paper in which actions against the unloaders were called senseless.[69] Abraham Kuyper commented that one could not blame the strikebreakers, who were driven by poverty to accept jobs in the harbor, but that their arrival would result in increasing unemployment and social disorder, which would require a permanent military presence in the harbor.[70] That Kuyper more or less sided with the people most hated by the dockers—scabs and soldiers—shows the gulf between the leader and his followers in the harbor. The Catholic papers *Maasbode* and *De Tijd* also ignored the stand taken by the Christian unions.

Dockworkers and weighers began a kind of guerilla action against "scabs" who cooperated with the Unloader Company. The manager of the corps of strikebreakers had to be protected against the crowd by the police; he was even fired at.[71] Since most seagoing ships contained parcels for more than one receiver, it often happened that on the same boat there were workers unloading grain by hand for German importers and personnel of the Unloader Company unloading by machine for Dutch importers. The scabs were continually harassed. As soon as the police came to protect them, the other workers would lay down their work. The Unloader Company would then proceed to unload the whole ship by machine. In the course of the summer, the violence between workers

67. *Patrimonium*, April 18, 1907.

68. Ibid., May 9, 1907.

69. Ibid., June 6, 1907.

70. *De Standaard*, June 6, October 26, October 30, and November 11, 1907.

71. *De Maasbode*, May 1, 1907.

and scabs increased. Rotterdam's mayor proclaimed a state of siege. Warships appeared in the harbor, and troops intervened in the fighting between the workers.

The German importers soon realized that their contract with the weighers was backfiring. They came to Rotterdam to negotiate with the Unloader Company, and in July all parties, including the workers' representatives, were trying to reach an agreement. The Unloader Company offered to raise wages, regulate working hours, limit the amount of grain transshipped by machine to 10 percent of the total amount during the next three years, allow weighers and other representatives of the importers to take part in the transshipment, and discuss the possible introduction of new unloaders with the workers' representatives (Spiekman 1907a: 749f.). The parties came very close to an agreement. But when the Unloader Company suddenly added a stipulation that the workers they had employed as strikebreakers be recognized as normal workers, and therefore be entitled to the same conditions of work and payment, the workers' representatives turned the proposal down (Spiekman 1907b).

The front of the employers on the side of the Unloader Company was now closing. The master stevedores, the Dutch importers, and the company started a fund from which to pay the weighers' corporation, to settle lawsuits, and so on.[72] When the workers wanted to negotiate again, the company refused to repeat its proposal on the ground that, since another firm was preparing to introduce unloaders, it could not afford to wait any longer.

On September 15 a strike broke out. The weighers did not take part in it; they had received another raise in pay from the German importers, and their contract forbade them to strike. The Christian unions, feeling that the possibilities of reaching an agreement had not been sufficiently explored by the workers' representatives, did not participate; however, there was a bitter note of resignation in their statements. They realized that they were too small to influence the course of events, especially since the employers were beginning to form a united front. The Roman Catholic newspaper most concerned with social problems, *Katholiek*

---

72. Gemeente-archief Rotterdam, Archief Graan Elevator Maatschappij, no. 55, "Contract van cargadoors ... etc." The failure of the employers in the first strike may be explained by the fact that they organized themselves much later than at other harbors, which has to do with the relatively small presence of liner companies and the prevalence of tramp shipping. See Broeze 1991: 178, 195 and Jansen 1979: 52f.

Sociaal Weekblad, wrote that new machines could not, "and in the interest of progress should not," be resisted—a grudging acceptance of the viewpoint of the leading Catholic ideologists mentioned above. It suggested, a bit late, that the municipal government exploit the machines, since an innovation with such effects on the working population should not be left to private enterprise.[73] Having rejected the strike, the Christian unions finally received support from their national superiors.[74] Like Abraham Kuyper in *De Standaard,* Roman Catholic papers emphasized the violent behavior of the strikers and praised the soldiers who kept a minimum of order.

The Algemeene Havenarbeidersvereeniging lent the strikers its support with some hesitation. Spiekman defended this standpoint. No longer speaking of abolishing the machines, he stressed the reasonableness of the wage claims of the workers in the light of the raise the weighers had obtained.[75] But other important SDAP leaders rejected the strike.[76]

While the conflict was running its course, the company was raising money to buy new unloaders. In its prospectus it stated explicitly that its goal was "to establish complete control of the transshipment and weighing of grain in the harbor of Rotterdam."[77] This time, the Dutch and German importers and the ship owners (the original shareholders) tried to secure as large a share in the new company as they could get (Cocheret 1933: 106ff.). Once the German importers had withdrawn their support, it was clear that the workers had lost the battle even before it had begun. Scabs recruited from all over the Netherlands and from Germany replaced regular workers.[78] Strike funds were far too small to

73. *Toenadering,* October 31, 1907; "Manifest. Aan de bootwerkers van Rotterdam!" September 23 1907, by Toenadering, Gemeente-archief Rotterdam, collection "pol. vakbonden alf." For the Roman Catholic standpoint see *Katholiek Sociaal Weekblad,* October 26, 1907, and July 2, 1910.

74. *Patrimonium,* October 3 and 17, 1907.

75. *Sociaal Weekblad,* September 28, 1907.

76. See also Spiekman's (1907b) account of this final phase in the conflict. Cf. Polak, cited in Cocheret (1933: 103) and *Voorwaarts,* November 23, 1907.

77. Gemeente-archief Rotterdam, Archief Graan Elevator Maatschappij, no. 2A, "Prospectus II." Cf. letter to the French engineering firm Terrin in Marseille, dated September 5, 1907, where the director of the Company asks Terrin to please hurry, "because of the present strike" (ibid., no. 2B).

78. *De Tijd,* October 3 and 14–17, 1907.

sustain a long strike, and on November 21 the strike was called off. The employers, by now united in a formal organization (the Scheepvaart Vereeniging Zuid), could dictate the terms of the peace. Five years later, there were sixteen unloaders, which together transshipped more than 90 percent of the grain that arrived in Rotterdam (Everwijn 1912: 641; Serton 1919: 39). Though in 1907 Rotterdam had lost some of its grain trade to Antwerp, after 1908 it recovered its share and outstripped other harbors. In 1909 Rotterdam received 3.5 million tons of grain (it had received 2.6 million tons in 1904),[79] Antwerp 2.9 million, Hamburg 2.3 million, London less than 2 million, and Liverpool 1.6 million (Serton 1919: 115f.). In spite of this growth, unemployment must have increased considerably, since the number of workers needed for transshipment had decreased by 32 percent (van der Waerden 1911: 222ff.) and since other sectors of work at the same harbor (such as the transshipment of ore) had also been mechanized; the recession of 1908 made things even worse (Serton 1919: 42).[80]

Conclusion

The introduction of the pneumatic grain unloaders at Rotterdam presented problems that were perceived and met differently by the various groups involved. The employers who introduced the machine were certain that it represented progress for the harbor as a whole. To them, the unloader was one more example of the inevitable and beneficial progress of technology, as liberal ideologists had formulated it. Their problem with the introduction of the unloader was one of control, in several respects. Beniger (1986) and Edwards (1979) have shown how the increasing scale and speed of production and transportation created a "crisis of control" in capitalist countries at the end of the nineteenth century.[81] The case of grain transshipment in a large seaport such as

79. *Het Volk*, November 21, 1905.

80. It is impossible to give an adequate account of employment in the Rotterdam harbor. The lack of statistics is accounted for by the fact that most dockworkers had casual jobs. See van Ijsselstein (1914: 7); see also "Buitengewone werkloosheid onder de havenarbeiders te Rotterdam, winter 1908/1909" [Report of the Burgerlijk Armbestuur to the mayor of Rotterdam about excessive unemployment in Rotterdam], Gemeente-archief Rotterdam.

81. Bloemen (1988: 90) says no such crisis occurred in the Netherlands. The story told in this article shows, I think, that he is mistaken.

Rotterdam is a perfect case in point. The complex network of the trade in grain came under severe stress as a consequence of the greatly increasing speed and volume of transportation that railroads and steamships allowed (Beniger 1986: 132–142, 172f., 249). Both employers and labor union leaders often described the process of transshipment as "chaotic."[82] But what looked like chaos was really a subtle game, full of tricks and bribery, that worked very profitably for some groups. The losers in this game—the ship brokers and exporters—tried to make it more profitable for themselves by seizing control of transshipment by means of a new machine. In one stroke, the grain unloader could eliminate factors and weighers (who manipulated the process in their own interest and in that of the importers) and many workers, thus lowering labor costs.

The immediate problem, however, was one of controlled introduction.[83] The directors and the shareholders of the Grain Unloader Company knew that the workers sometimes reacted violently to the introduction of new machinery. The situation was complicated by the fact that the employers did not form a united front: factors and master stevedores would certainly oppose the unloader. The fact that at first only two machines could be bought made the enterprise more vulnerable still. The Unloader Company therefore tried to make the transition to mechanized transshipment a gradual one, by fitting the machine into the existing system—that is, by making it acceptable to the master stevedores and the factors. The workers would simply be forced to accept the new situation.

Once introduced, however, the machines provoked fierce resistance from those who profited most from the existing system. The importers, in particular, were a powerful enemy that the Grain Unloader Company had not reckoned with. In the course of the conflict, the company came

82. The German importers also complained of the Rotterdam harbor in these terms; see *Rotterdamsch Weekblad*, May 18 and July 20, 1907.

83. It is interesting to compare the introduction of the grain unloaders to that of ore transshipment machinery, around the same time, which hardly provoked resistance. The critical differences seem to be the fact that Van Beuningen, who was responsible for introducing the machine, managed to keep this introduction a complete secret until the day it was put to work. In addition, Van Beuningen's familiarity with his workers may explain why he succeeded in dissuading them from striking against the machine, and labor relations were much less complex and therefore more manageable than in the grain sector. See Gemeente-archief Rotterdam, manuscript 326, autobiography by Van Beuningen, pp. 27ff.

to realize that it had to establish complete control over the transshipment of grain, since only this would convince the importers and silence the workers. This was done by introducing more machines.

It is much harder to know the workers' opinions, although those of their leaders were amply published. The desperate determination of their actions (especially the long strike of largely unorganized workers, without strike funds, in the fall of 1907) indicates that they saw the unloader as a lethal threat, likely to deprive them of their daily living.

In the debates between local dockworkers' leaders on the one hand and national politicians and labor union leaders on the other, we can see that the way modern technology had been "intellectually appropriated" by the elite was contested at the local level by people whose daily lives were threatened by the coming of a new machine. National leaders routinely repeated the general consensus about the inevitability of technical progress and the need to create conditions in which it would be profitable for all. To them, trying to block technological change was both reactionary and hopeless, and the consequences of mechanization, such as (supposedly temporary) unemployment, simply had to be accepted. However, local leaders, both socialist and Christian, felt that taking this position came down to abandoning their men, with the very probable consequence that the men would abandon them. They could not share the stoic acceptance of unemployment and poverty that the dominant discursive framework implied. Therefore, they tried to legitimate the actions against the unloaders in terms of the dominant ideology, in an effort to persuade their superiors. Thus, Spiekman argued, the machines could not be seen as serving the interests of the workers or even those of the harbor. Toenadering used Abraham Kuyper's theories of a corporate society to claim that workers had to be consulted on mechanization—a view that was shared by "Kardinaal Manning." Sam van den Berg emphasized brotherhood and community as the basis of socialism and denounced the technological determinism of the Social Democrats, suggesting that this dogmatism was similar to that of the churches in that it aimed at creating a meek fatalism among the workers. Van den Berg believed, however, that "the power of facts" and a deep sense of justice had defeated ideological determinism, and that the workers were not to be fooled by abstract theorizing (van den Berg 1906: 30f.).

None of the local leaders rejected the new machinery out of hand. Their main objection was to the way the unloaders were introduced.

And here an idea appears that we have not encountered in the above discussion of the ideological statements of the different groups: the idea of the harbor as a community and of the unloaders as intruders.[84] There was a basic sense, especially pronounced in the speeches of Sam van den Berg and in the publications of Toenadering and the Roman Catholics, that the employers had no right to impose a machine that would deprive hundreds of workers of their daily bread.[85] Some, including van den Berg, Spiekman, and (at first) the Catholics, denied the economic necessity of the unloaders without rejecting labor-saving machinery in general. Van den Berg was unique mainly in that he thought that *some* machines, such as the grain unloader, harmed the interests of so many people that they should be permanently abolished. This idea of community may explain the vehemence with which violence (or accusations of planning violent action) was rejected, even by radicals like Sam van den Berg's brother. The only violence that occurred was directed against men imported from outside Rotterdam to break the strike. Feelings against these men were exceedingly bitter; they were manhandled and thrown overboard during the fights in May 1907, and an impending agreement in the summer of 1907 failed essentially because the employers refused to disband their corps of strikebreakers. Bitter complaints (as in the *Elevator-lied*) were also directed at the shareholders in the Grain Unloader Company that came from outside the Rotterdam harbor community.

The workers' leaders knew very well that the Rotterdam harbor was not the community it should be. They often described it as a jungle in which only the profit motive counted.[86] But they believed in the possibility of creating ''order,'' a situation in which everyone had his recognized and protected rights. Their idea of a well-ordered community was the mirror image of the employers' search for control, and it led, much later, to more regular labor relations, mediated by employers' and workingmen's associations.[87] This did not imply a change in views of the role

84. The word ''intruder'' occurs on p. 26 of Berg 1906: 26.

85. See also a pamphlet addressed to the Rotterdam grain workers by the combined workingmen's associations in August 1905, written by Wessels and cited by Berg (1906: 10).

86. E.g. *Toenadering*, February 3, 1906, August 22, 1907, and October 31, 1907; Spiekman 1900: 122f.; Mol 1980: 178. The first victory in this campaign for order was the introduction of a labor inspectorate in the harbor, started in 1906 after a desperate plea of the union leader Wessels.

87. Rotterdam therefore moved with the general tendency, described by Broeze, from militant radicalism, via syndicalist failure, to reformist pragmatism (Broeze 1991: 186).

of technology in labor relations. On the contrary: politicians and labor union leaders saw in the defeat of the strike a confirmation of their technological determinism. The idea that the introduction of new machinery should be subject to negotiations among all parties involved disappeared, only to be revived during the 1970s with the introduction of containers (Nijhof 1990: 28). The workers themselves never resisted the introduction of new machinery again.

9

Sociological Reflections: The Technology Question during the First Crisis of Modernity

Peter Wagner

Technology and Modernity

We are all used to speaking about "modern technology." The marriage of the terms "modern" and "technology" has been so widely used as to make this linguistic couple almost inseparable. If "modern," however, is supposed to mean more than just "most recent," one must be able to say what keeps the two together. A possible response to this question is that to call technology modern means to adopt a rationalist and instrumental view and to focus on the functional efficacy of technical means with regard to the enhancement of human capacities. Indeed, one may reasonably argue that the widespread use of the term "modern technology" reflects the dominance of such a conception of technology throughout much of the post-World War II era, at least until the late 1960s.

At first sight, the linkage of "modernity" and "technology" seems perfectly appropriate, in that the history of the West, if reconstructed as the "project of modernity" from the scientific revolution and the Enlightenment to more recent achievements such as space travel and the manipulation of the human genome, can easily be described as the striving for human autonomy and the mastery of nature. Furthermore, such a view appears to be confirmed by the social history of technology. For instance, cars and airplanes, which have allowed us to enhance our control over physical space as much as they have enabled us to move about individually at greater distances from the social spaces in which we spend most of our everyday lives, have been regarded as "heralds of modernity" since the first decades of the twentieth century (Gentile 1988: 108; Overy 1990).

Nevertheless, I shall argue that "modernity" does not translate easily and unequivocally into "technology," and that the rich reflections about the relation of human beings to technology produced in the first three or four decades of the twentieth century can be taken as evidence of a much more complex relation between those terms. Not only are these reflections as diverse as they could be, reaching from unconditional praise (as in futurism) to gloomy perspectives on the ultimate subjection of human beings to the exigencies of machines over which we have lost control; they also fill a much broader space of possible discourses on technology than the rationalist and instrumental conception associated with "modern technology."

The authors of the preceding chapters have demonstrated the variety of views on technology that were held in the various national contexts. I shall try to show how this variety of debates, in sum, widened the space of the technology discourse in comparison to the nineteenth century, and how this widening took place in the double context of rapid diffusion of new technologies in society and the broader socio-political constellation in which European societies found themselves in the years around World War I, a period I have described as the "first crisis of modernity" (Wagner 1994). The rationalist and instrumental conception of technology emerged as dominant from these intellectual struggles after World War II. However, rather than seeing this historical outcome as the settlement of an issue by finding the superior solution, one should regard it as a temporary closure of a debate that went along with a socio-political transformation that I call a "closure of modernity."

The present chapter complements, and does not contradict, the preceding chapters. All technology debates developed in particular socio-political contexts which they addressed and tried to reshape; such contextual interrelations can only be explored in detailed case studies. However, the intellectual and socio-political world of the early twentieth century showed interrelations and interdependencies between local and national situations to such an extent that not only is a cross-cutting analysis possible; such an analysis is necessary if a broader picture is to emerge. The argument of this chapter remains, nevertheless, historical, and does not entail any assumption of suprahistorical logics or dynamics.

To show how a situation was arrived at that demanded a new approach to technology toward the end of the nineteenth century, I begin with a brief sketch of the views of technology that were dominant during that century. This new approach emphasized order and control. Technologies were increasingly set up as technical systems covering social spaces and channeling human action into preconceived corridors. At the same time, political debate distanced itself from classical liberalism and focused on collectivities whose interests were seen as predetermined. Partly as a reaction against these developments and partly as an independent proposal to deal with the issue of technology, conceptions emerged that related technology directly to key questions of human existence; this is what has often been described as the response to a dramatically new experience of space and time. World War I led to the collapse of whatever vague consensus there may have been before and produced a divergence of views on social and technical developments. Between the wars, the contours of the new, "modern" approach to technology emerged, particularly forcefully in the United States; they were transferred to Europe strongly only after the end of World War II. Most recent technology debates, however, show that this approach meant nothing but a temporary closure of an issue that will re-emerge, or perhaps already has re-emerged.

The Social Imagery of Technology during the Nineteenth Century

The social configurations of the first half of the nineteenth century were radically divided in their views on technology. After the onset of what has come to be called the Industrial Revolution, the invention and introduction of new, powerful technologies for production and transportation was widely hailed as a triumph of the creative power of man. Even during that period, though, there was no lack of critics. "The machinery question," as Berg (1980: 3) has suggested, "stood at the center of the stage of social, political and intellectual conflict in the early nineteenth century." (See also Mithander 1991.) Romanticism is often regarded as the first literary-intellectual reaction against the spread of industrial technology, and "The Sorcerer's Apprentice" and Goethe's version of the Faust story are early examples of the recurrent "technics-out-of-control" theme in social thought (Berman 1983; Sieferle 1984; Winner 1977)

More specifically, there is a social division over industrial technology. "Creative man" was incarnated in the figure of the bourgeois entrepreneur. That he was a property owner and that he was male were widely seen as preconditions of an existence as a responsible, active, creative individual. In contrast, many workers and their families witnessed the introduction of new technologies in connection to the deterioration of living conditions in general and conditions in the workplace in particular. "Luddism" became a generic term for the radical rejection of new technology. It was counterpoised to the conviction that technological development as such marked progress—a view that can be called the dominating perspective among established intellectuals throughout most of the nineteenth century.

Marx and Engels's *Communist Manifesto*, published in 1848, remains one of the most telling accounts of the transformation of a social configuration by a specific social class and by means of technology. No great interpretive effort is needed to read a deep fascination for the transformative power of the efforts of this class-*cum*-technology into Marx and Engels's account. At the same time, *The Communist Manifesto*, together with Engels's 1845 essay "The Conditions of the Working Class in England," marks a transition in the thinking about technology. It offered those who appeared to be suffering from the aforementioned transformation a perspective beyond mere rejection of the machine. Marx and Engels suggested that there were potential benefits associated with machines, but that the technological advances had to be socially appropriated by a radical restructuring of society. Upholding the radical critique of the class that had invented those means and had first put them to use, they did not view technology as responsible for all evil. The "development of the productive forces" meant progress indeed in the Marxian view; however, the social utility or damage of technology depended strongly on the ways it was used (not to speak of the fetters that bourgeois relations of production imposed on the productive forces).

Without wanting to overemphasize, it is significant indeed that "the machinery question" of the first half of the nineteenth century was succeeded by "the social question" in the second half. Political economy (the discourse on society that reigned supreme among the commercial classes) and its socialist critique shared the view that neither technological nor economic rules and rationalities as such were problematic but that the social context that governed their workings and effects was. Thus, the middle of the nineteenth century marked a point of comparatively high acceptance of technology (Mithander 1991: 11). During the following decades, political and intellectual attention was refocused onto issues of social reorganization, this debate increasingly being structured by the opposition between established liberal-bourgeois and aristocratic elites and the organizations of the workers' movement in many countries. At the same time, this period was marked by strong industrial development, particularly in the United States and Germany, and by what later came to be called a new wave of technical inventions and a second industrial revolution.

This is not the place to describe the effects of these changes on the social configurations in any detail or to repeat accounts of the dislocation of large parts of the population, of the growth of industry and the emergence of industrial cities, or of the formation and the rising strength of the workers' movement, its parties, and its social theories in the second half of the nineteenth century. Suffice it to say that these processes, reordering social practices and disembedding individuals from the social contexts in which they had grown up, uprooted social identities and created widespread uncertainty about individual life chances. Many of these developments appeared to have links to technology, and they tended to alter views of and attitudes toward technology again, even if only gradually. Among the elites, at least among the more prudent and conservative parts, doubts grew as to whether they were not themselves growing the seeds of disorder and instability into a social configuration which they dominated. "Mechanical invention," Overy (1990: 73) notes, "was an important part of the bourgeois-liberal world-view. . . . Yet the same bourgeois idealism that gave rise to the striving for order encouraged a progressive scientific and technical culture which threatened disorder. . . . In a world of increasing uncertainty, the disintegrative impact of radical invention was all the greater." This quotation refers to the extension of the reach of human action as an effect of technology. Though desired, this effect is disturbing since it tends to shake established conventions and increase contingency in social life. Also during the closing decades of the nineteenth century, however, a different view on the possible use and effect of technology gained in importance, not least among the engineers, entrepreneurs, and administrators who were actively engaged in putting technology to use. It was recognized that technical means could be deployed in such a way as to structure and organize an entire set of social practices. This means that materially predefined rules of action would be extended to a social space so as to cover it—or one kind of actions in it—completely. Not extension, but control of human action is the main effect of technologies deployed in such a way, and the intention is to decrease contingency and enhance predictability.

Though any brief categorization will tend to be an oversimplification, I might dare to say that the prevailing perception of technique early in the nineteenth century was that of individual machines or factories, which, though they were growing in size and power, were handled, supervised, and controlled by individual human beings, be they workers or entrepreneurs. Toward the end of the century, perhaps starting with

Marx (Berg 1980: 340f.), this view had clearly given way to one that emphasized the technical, economic, or social connections between the single elements and, thus, the functional (or, as some would say, in the long run dysfunctional) interrelatedness of those elements as parts of a system. This new view contained elements of a way of reappropriating technology, though they were not yet realized during the nineteenth century. After the innocent conception of harmonious technical progress was shaken, a view of technology as a mechanically or systemically integrated large system could be a means of restoring order, or, rather, of bringing about a new and better order. Inventors and innovators sometimes saw themselves as builders of new systems that would put society itself on a different footing and would point to ways out of the dilemmas that rapid and uncontrolled industrialization had produced (Hughes 1989, chapter 5).

The Building of Technical-Organizational Systems

To grasp the background to these changes in debates, let us briefly look at two aspects of the social transformations during the later nineteenth century that are closely related to the development of technology: the restructuring of allocative practices that has come to be labeled the Second Industrial Revolution and the emergence of the oligopolistic managerial enterprise. The former refers, in particular, to the applications of electricity and chemistry, to the new forms of transportation based on the internal-combustion engine, and to the telephone (Landes 1969; Hughes 1989; Radkau 1989). Organizational changes went along with material changes. The years before and after 1900 witnessed the emergence of the large-scale, increasingly managerial enterprise and of the proposals for the planned restructuring of the production process that came to be labeled Taylorization (Piore and Sabel 1984; Noble 1977). I shall not reiterate familiar themes of these transformations here, rather, I shall focus on certain features of these reorganizations that are essential to understand the perceptions of technology.

As I briefly indicated above, the late nineteenth century has often been described as an era of technical innovation. The first two quarters of the twentieth century, in contrast, were mostly characterized by the permeation of society with technologies invented earlier. Elsewhere I have argued in more detail that this distinction of periods, if valid, can be related to a long process of social sedimentation of innovations which occurred as a part of a stabilization of an entire socio-economic paradigm. In this perspective, the so-called late-nineteenth-century wave of innovations was related to a major transformation of social practices, which I have labeled the transition from restricted liberal to organized modernity (Wagner 1994: 77–81).

Concerning the uses of technology, this transition can, very schematically, be characterized as a movement from an emphasis on the extension of the reach of human action to an emphasis on the control of social and natural spaces. The reach of action over spaces was to be controlled by establishing a material connection or by strengthening the chains of unambiguous interaction. In principle, interaction chains had assumed global extension since the era of the discoveries. With nineteenth- and twentieth-century means of transportation and communication, however, long interaction chains acquired more of a routine character, and they were much more standardized. Modern institutions often established unambiguity of interaction from the beginning and upheld it all the way along the chain. They did so mainly by two means. First, they brought the information or good that was handled into a shape that was transportable with the technique to be applied. This process meant work on the good that entailed its reduction to some basic characteristics. Storage and transportation possibilities, for instance, altered food markets and allowed their delocalization. Economies of scale then led to the mass-produced foods that have become typical of the twentieth century. Second, efforts were made to ensure that the interaction chain was closed to interference from the outside. Among the material means of such closure were iron tracks for the railway system, wires for the telephone system, and, later, concrete roads for the highway system. By such means, the reachable distance was extended considerably; at the same time, permissible paths and access points, and the micro-behavior of individuals inside the system, were rigidly prescribed, and communication and renegotiation about the rules of such behavior were virtually precluded. Furthermore, a boundary was erected between those inside and those outside; for example, the inhabitants of a village that lacked an access point to a rail system or a highway system were further removed from the other members of society than they had been before.

Key features of such material practices are their simplicity and their independence. By the former, I mean that they destructured more complexly related interactions and recomposed them in preconceived, more orderly and predictable ways, often appearing as uniformity once such a practice had become socially dominant (Radkau 1992: 13f.). By the latter, I refer to their delocalized character. Such technologies are conceived to be universally applicable once certain minimal requirements are met. Being simple and independent, they could take the form of "systems" that could be imposed on a local field of action. In these terms, the second half of the nineteenth century and the first half of the twentieth can be characterized as a period in which social space was (literally) perforated by technical networks: the railroad, the telephone and electricity networks, car-usable road networks, radio and television broadcasting systems. The growth of these "primary" technical systems was always based on economies of scale in one sense or another. The cost of building and/or maintaining a system could be so high that it could be run cost-efficiently only with mass usage. Or a system would be attractive to users only if it had wide coverage, as do the telephone network and (for advertisers) the broadcasting systems. And a wide extension of such networks would provide growth paths for producers of equipment to use them, such as cars or electrical appliances. All these features entailed a move toward standardization of products and homogenization of patterns of behavior.

A particular example of a social technology showing similar features is the reorganization of production through "scientific management" (later called Taylorism) and, related to it, the assembly line (associated with the name of Henry Ford). Two main reasons are usually given for the social attractiveness of Taylorism: that it increases efficiency and productivity and that it expropriates the workers not only of their skills but also of their control over the work process. Naturally the first argument tended to be used for the employers' side, the second for the side of the workers' movement. I would like to stress a more general aspect of Taylorism. With its minute decomposition of human movements, scientific management ended up reducing every action into a limited number of component parts. The production process could then be reassembled from these known and measurable parts. In theory, scientific management had complete knowledge of the labor process at its disposal. Such knowledge could then be used for a variety of purposes, whether they were called "efficiency" or "expropriation of the workers' power." At its basis, however, was the establishment of order and certainty, stability and predictability, on a recalcitrant reality in the factory.

These uses of technology went along with the appearance, in the last half of the nineteenth century, of "a new form of capitalism . . . in the United States and Europe" (Chandler 1990: 1; see also Kobayashi and Morikawa 1986). Toward the end of the nineteenth century, the average size of a firm had grown sharply, partly through direct organizational expansion and partly through mergers. The emergence of the modern, big business enterprises may be related directly, as Chandler does, to organizational requirements for managing new technical systems, such as railroad and telegraph systems. Subsequently, then, the existence of this new organizational form allowed and stimulated mass production and mass marketing. It will not suffice, however, to point to technical innovations as the main cause for the growth of firms, since some of these techniques showed a long maturation period before they were widely applied. The growth of firms, though, can more precisely be located in time as following on the long depression at the end of the nineteenth century (1873–1895). Organizational expansion can be analyzed as an escape from the vagaries of the markets under competitive capitalism.

If the market share of a firm's product is increased, the possibility of controlling the market is enhanced. All economic theorizing that focuses on automatic equilibration and maximization via markets has to assume that economic actors are exposed to the workings of the market without being able to strategically shape it. Big firms, however, establish a new kind of economic agency when they are able to influence the conditions of market exchange owing to the size of their own share. Through this kind of organization, companies do not merely benefit from economies of scale, if narrowly understood in technical and economic terms, but they produce a social advantage, namely manageability on their own field of action.

Later in the twentieth century, since the modern business enterprise acting on oligopolistic markets has become the dominant type of firm, it has repeatedly been argued that the development of advanced capitalism is much less associated with market competition and dynamic entrepreneurship than with increasing organization of production and distribution. The move toward organization should indeed be seen as an attempt to control conditions of action in a general context of fluidity and change. "Risk avoidance and organizational stability," notes Lehner (1983: 439f.), "is the usual device of large organizations and firms."

This reaction on the part of "capital" is thus not so much unlike the parallel one on the part of "labor," namely to organize a share of the market (for products or for labor) as big as possible so as to control it instead of being exposed to it (Offe and Wiesenthal 1980). It is a move to re-establish certainty under conditions of great uncertainty. Following the principles of bureaucracy, big organizations try to cover as much of the relevant field of action as possible, and to structure their actions on this field according to clear and fixed, hierarchical rules.

Such ways of elaborating and using technologies (material as well as social ones) are part of a major change of a social configuration, a strong shift in emphasis among some of the main principles that orient action. Depending on the analytical emphasis, there are various ways of describing this shift. One may talk, as above, about a shifting direction of technology use, from extension of reach to control of space. In a very similar way, public debate about societal reorientation at the time referred to a shift from external to domestic, or interior, colonization. (See chapter 6 above.) The increasing emphasis on organization and predictability is also expressed as a turn away from individualism toward collectivism.

However, the broad historical pattern of technology use that I try to carve out should be read neither as an argument for functional superiority of the new order nor as a strong scheme superimposed on diverse and conflict-ridden social realities. The building of technical-organizational systems was a historical response to uncertainties and risks in nineteenth-century capitalism, both on the increasingly international markets and in relation to the increasingly assertive workers' movement. New forms of control were indeed established, again in both directions. However, they also contained the germ of new problematics that should come fully to the fore in our current restructuring of capitalism.[1]

And beyond an enormous variety of national developments, which limits the possibility of general statements, it is also necessary to point to technologies that deviate from the historical pattern—though less in their character of being part of systems than rather in the forms of usage. The car and the telephone are examples. Historically, the diffusion of both technologies falls squarely into the period at issue here, the car as a product being even the prime example for the emergence of an "organized" production and consumption pattern. However, the forms of use that both allow may be highly individual and private, and were early on recognized as such by the users. In both cases, early restricted patterns of use, such as for military and for business purposes, were soon exceeded, and these techniques became the symbols of independence, autonomy, and individuality. The car-and-road system even tended to supersede and replace a transportation system that was much more

1. For elements of an assessment in those terms, see chapter 8 of Wagner 1994.

collectively arranged, the railroad. The specific character of such inventions and their uses occasionally garnered them a special position in the debates on technology. Furthermore, the de facto (though mostly untheorized) recognition of their difference made it possible to break up the question of technology, to remove it from a position where the answer to it would determine the fate of society and humankind to one where there could be a variety of technology-specific answers.

The technology debates of the early twentieth century can, in general, nevertheless fruitfully be analyzed in the terms introduced above, in terms of a dominant shift toward organization, collectivity, and control. This shift, however, was accompanied by different, partly directly opposed views emphasizing autonomy, individuality, and freedom. The breadth of technologies the wide experience of which was relatively new offered examples to support the former as well as the latter view. For this very reason, those technology debates have opened a very wide space of possibilities to talk about technology. In the following sections, I shall try to reconstruct this discursive space.

Technology and the End of Liberalism

At about the time when technologies became increasingly perceived as interlocking, systemic arrangements that could serve to order and control physical and social space—and less as single tools enhancing human power for freely chosen purposes—the predominant view on the relation of individual human beings to the social configuration shifted in an analogous direction.

Generally, the latter half of the nineteenth century can be analyzed as a period of a decline of the hegemony of classical liberalism. The prevailing political view of the middle of the century saw the liberal individual as an active part of a national setting in which he, among other things, used technologies to realize his self. This world view insisted on the autonomy and responsibility of the individual, even if the extension of this view was typically restricted and women and workers largely excluded. Social movements of the latter half of the nineteenth century demanded full inclusion of all individuals under such conceptions. Faced with such demands and unable to reject them, classical liberals saw the emergence of a mass society in which the active attitude of the bourgeois individual would be undermined.

Until about the time of World War I, broadly liberal positions maintained a stronghold among members of the intellectual establishments.

Socialist reasoning was growing, but either it developed close ties to liberal ideas and milieus, as in England and France, or it was kept off institutional positions of importance, as at German universities. However, liberalism had lost all the enthusiasm its promoters had shown through much of the nineteenth century. In many cases, awareness had grown that the basic concepts of liberalism had to be strongly reconsidered; classical sociology, for instance, is marked by such views (Seidman 1983).

In the perspective outlined above, much of the history of the European nineteenth century can be read as an increase of the awareness of contingency (Berman 1983).[2] Toward the end of the century, this feature of the "modern condition" provoked strong efforts at decreasing such contingency. As a political theory that kept many issues open and contingent as a matter of principle, and as a political practice whose effects were often seen to be dissolving traditions and certainties, liberalism was attacked from two angles. On the one side, the commercial elites who had, to a large extent, supported classical, restricted liberalism came to recognize that their own practices, if continued without control, would undermine the foundations of the social order. On the other side, the social movements that demanded inclusion did so in the name of collectivities, rather than individuals, and were accordingly open to redefinitions of society as a collective order.

Such collectivist options, of which there were many discursive variants as well as a number of different ways to inscribe them into laws and institutions (Ewald 1986; Rabinbach 1994), often drew support from an analysis of technology and economy that held that the industrial mode of production had changed social relations to such an extent that individuals could no longer be regarded as exclusively responsible for their actions, since they had come to occupy predefined places in a larger, functionally related order. A view emerged emphasizing that humankind was now exposed to technical dynamics; that masses were being organized in rationalized, homogeneous orders; and that the individual was lost. To such a view corresponds a machine image of society, which— although the idea is much older—gained increasing acceptance during the early decades of the twentieth century (Marz 1993).

Max Weber's use of the "iron cage" metaphor is a significantly cau-

2. The term "increase" should not be read in linear, much less evolutionary, terms. As Toulmin (1990) argues quite convincingly, historical periods of rather great certainty may alternate with periods of awareness of contingency. Any such sequence is tied to events rather than to a direction of history.

tious example of such a conception. Though Weber applied this term specifically to bureaucracy, he clearly had a view of the formalization and rationalization of human action more broadly understood. Indeed, by Weber's time Oswald Spengler had already fused "the metaphorics of the machine with the rejection of liberalism" by arguing that the exigencies of technology demanded to put an end to soft-spoken liberalism (Maier 1970: 44). Later critics of the mass society, including Max Horkheimer, Erich Fromm, and Hannah Arendt, would similarly elaborate on this metaphor and, turning it negatively, would speak of human beings as "cogs in the machine" of industrial mass capitalism—of the subordination of man to the rhythm and logics of the machine.

Theorists of society as a machine detected a new rationality in the emerging social configuration, whether they welcomed it or deplored it. This new rationality, in a view such as Weber's, entailed rationalization in the sense of a loss of communication and consensus about ultimate values—the well-known "disenchantment" of the world. Weber tried to weigh the enablements brought by such machine organization against the constraints, and interpreters have remained undecided as to the conclusions at which he personally may have arrived. Certainly he saw in the enablements one of the causes for the historical emergence of this phenomenon, since human beings chose it for its advantages as to the possibilities for rational action. In the views of critical theorists, in contrast, the rationality is always socially more one-sided; it is seen as the rationality of the ruling classes and as the logics of the capitalist order. Both interpretations, however, are basically as rationalistic as the machine metaphor may, at a first look, suggest.

For others, however, it was not least these technologies of the modern age that brought a kind of reenchantment that was directly linked to the suprahuman powers of the machine. Generally, such thought was not alien to Weber, much of whose work can indeed be read as a personal struggle, which remained undecided until his death, between a rationalistic and a Nietzschean reading of modern times (Peukert 1989). However, no intellectual of Weber's age, having grown up with the nineteenth-century debates, would link the desire for enchantment positively to machines. However, some authors of Weber's time, but of a much younger generation, went in this direction, and fell into a much different tone. Their writings provide us with the possibility of pointing to ways of appropriating technology that have not become dominant in those configurations of modernity that have developed historically but which may remain relevant for understanding the relation between human beings and the objects of their creation.

The views on technology voiced during the early decades of the twentieth century can fruitfully be analyzed, as I will try to demonstrate, as varieties of ways of dealing with a historical perception of a radical increase of contingency during the nineteenth century. All these ways show the inescapability of the condition of modernity, since they can be read as variations of the double imaginary signification of modernity, of the ideas of autonomy and mastery. However, they do indeed display remarkable variations of this theme. Some strengthen the individual understanding of autonomy, whereas others propose and enhance a collectivist reading of this term; some intend to extend the instrumental mastery of the natural or of the social world, whereas others intensify possibilities of mastery of the self. In many readings, specific technologies such as were developed and used early in the twentieth century were an important ingredient in elaborating and supporting the particular view of the "modern" world, namely in establishing a viable linkage between autonomy and mastery. The possibilities of dealing with contingency that were being outlined in this debate have established such specific linkages. They have marked a field of possibilities of dealing with technology in which we still move, though significantly not very close to either of the extremes.

Technology and Revelation

The turn of the century was characterized by the feeling that the nineteenth century had come to a close not merely in temporal terms but that it had exhausted its energies and moved into a deadlock. While classical liberalism had clearly been superseded, the collectivist arrangements that had increasingly been introduced in its stead were sometimes not seen as providing an alternative, rather as stifling cultural potentials. This mood prepared the way for a new liberation in aesthetic terms that was not least inspired by technical developments (Kern 1983; Harvey 1989, chapter 16).

Among such currents, Italian futurism is best known for its unmitigated praise of technology. In the terms of Emilio Gentile (1988: 107), futurism developed a mythical image of modernity as an explosion of human and material energies. Here the machine, far from being an iron cage, enables human beings to grow beyond their own physical capacities and to reach higher regions of being. Without doubt, there is a tragic element in such views, since humans are reduced in size and importance and cannot escape this situation. At the same time, though, they have shown their greatness by creating exactly those objects and by merging with them to live new kinds of experience.

According to futurism, as in many other views of the time, technologies have outgrown human beings. Rather than an ally, technology has become the master of humankind (Nazzaro 1987: 78). It teaches and disciplines human beings. "The machine gives lessons in order, in discipline, in power, in precision and in continuity," as Filippo Marinetti said in 1924, already in the context of fascism (Masini 1988: 309). It even demands the ultimate sacrifice: "Blood is the oil that the wheels of the machine need that flies from the past into the future," Giovanni Papini wrote in 1913 (Gentile 1988: 114). But technology does not simply subjugate human beings; it elevates them to greater heights, which are not accessible without it.

In this sense, it is indeed appropriate to analyze futurism as a "religion of technology" (Tessari 1973: 209). Technology is a higher entity that is praised because it provides revelation and redemption. More generally, we may speak here of a metaphysical conception of technology that certainly was not specific to futurism. Technology comes to be seen as something bigger and longer lasting than an individual human being, who, in turn, becomes part of that structure and fits into it. This is a position that hitherto was reserved for openly transcendental phenomena—such as God, nature, and reason—or for highly valued sociohistorical phenomena defining individual identities, such as the family, the nation, and—as an intellectual creation of the workers' movement—social class.

In Germany, Ernst Jünger developed a related cosmological vision around the experience of industrial work. In contrast to the futurist praise of the machine, however, for Jünger machine and man together develop a new form (Gestalt) of higher, "existential" being that overcomes the limits and contradictions of bourgeois society: "Technology is the domination of a language which is valid in the domain of work. This language is no less meaningful, no less profound, than that other sort which belongs not only to grammar but to metaphysics. Here the machine just as much as man himself plays a secondary role. It is only the organ through which this language will be spoken." (Jünger 1932: 150; cf. Orr 1974)

Jünger insisted that this Gestalt is undescribable, since it has no qualities and no inherent values. One may read him as trying to develop a worldly metaphysics: no substantive transcendental elements are given, but the idea of an ultimate orientation in life beyond mere existence is maintained, an orientation which is to be found in the form of the interaction of machine and man. The peculiar force of Jünger's writings stems not least from the fact that he couches in mythical, heroic language what otherwise could be read as a sober praise of functional efficacy. Where Weber struggled with the peculiar irrationality of rationalization without substantive objective (Löwith 1932: 41), Jünger turned technical rationalization back onto itself and elevated it to higher meanings.

Like Jünger, Martin Heidegger (1982: 12) saw humankind in an unprecedented confrontation with technology in the 1930s, a confrontation that would change the human condition entirely. The basis for Heidegger's thought was the insight that technology that had outgrown human beings was "a way of revealing" (eine Weise des Entbergens) rather than merely a means to something. Rather than being placed in a frame of human ends and purposes, technology enframed human action. That is why any merely instrumental and anthropological perspective on technology would turn out to be unsatisfactory; in these terms Heidegger (ibid.: 23f. and 6) indeed marks the core of the technology debate and its post-World War II outcome. During the 1930s, Heidegger thought that humankind might live up to this challenge—"the ominous frenzy of technique let loose [and] the rootless organization of standardized man" (1959: 37), as visible both in the United States and the Soviet Union—and rise to a new condition. But after Nazism and World War II, Heidegger had given up any concrete hope[3]; now he focused on the arts as a way of questioning technology.

In a short essay such as this, there is no way to do justice to the subtleties and complexities of conceptions of technology in those debates of the early twentieth century—nor is this the main ambition here. However, the sketch of positions given up to this point allows us to map the discursive space of possible appropriations of technology as it was constructed during this period.

Technology, Man, and Society in the Early Twentieth Century

The very schematic discussion of views on technology in the nineteenth century had left us with two basic positions. Either (as in the bourgeois progress perspective) technology was seen as liberating human beings

3. See Heidegger 1982.

from the limitations of their natural endowments and enhancing autonomy and mastery of the world, or technology itself needed to be liberated
from its bourgeois appropriation to unfold its full beneficial potential.
The latter view added an important twist to the argument but did not
substantively change it. The two views had in common that they did not
discuss, much less question, what substantive effects technology actually
had.

The liberating effects of technology were widely seen as problematic
(again) by 1900, and especially after 1918. From the late nineteenth
century on, the substantive discussion of technology was a part of the
opening of the restricted liberal order of social practices that I analyze
as a first crisis of modernity and an increased awareness of contingency.
The technology-related parts of the politico-intellectual debates of the
time are especially significant since technology is often considered in a
double way: as a source of uncertainty and uprooting on the one hand,
and as the key to the reestablishment of certainties and the re-embedding
of social identities on the other.

A common theme of this discussion was that technology, in view of
its immense transformative powers, could hardly be seen as existentially
neutral, or even as neutral in the broader sense of generally enhancing
human capabilities. The individual and collective experiences and interpretations implied that technology had another, deeper meaning. The
degree to which technology was endowed with existential meaning is
one major criterion by which views of technology can be distinguished.
The other criterion is the question of what one might call the social
interlocutor, or addressee, of technology. As some authors put it, technology speaks to human beings, raising questions and demands. The views
vary significantly on whether it speaks to individuals, to humankind, or
to certain more or less well-defined collectivities. It is along the lines of
these two criteria that I shall map the discursive space of technology
appropriation.

Quite a number of the intellectual appropriations of technology
endowed it with revelatory and redemptive powers. There can be little
doubt that the emergence of such views is related to immense and
hitherto unknown transformations of human experience as they were
effected by some of the new, or only recently diffused, technologies.
The car, the cinema, and the telephone restructured experiences of
space and time; the big factory, the world war, and the city provided
new senses of human social organization and interaction. These themes
are taken up in artistic as well as in intellectual expression. Still, these

technical themes acquire their fundamental significance not really on their own, but in the context of the uncertainties brought about by the broader social transformations of the time. And the general perspective leaves the normative appreciation still open. Authors may write in very similar terms and style whether they are abhorred or positively impressed by the workings of technology. They may praise a new, higher order of human life on the horizon of the future, or may see the end of humankind approaching; in both cases it is technology that provides revelation.

Beyond (or perhaps between) such diverse evocations of revelatory aspects, what can be called a classic modern view of technology emerged. Without dealing with technologies in the material sense of the term, Max Weber's theorem of rationalization provides the foundations for such a view. And within the realm of socio-philosophical discourse, Georg Simmel's writings on money can be seen as one such attempt at a "sober" analysis of technology as the formalization of human action and its extension over wide spaces and across time.

This classic modern position tries to deal with technology by focusing only on instrumental, procedural aspects, and avoiding any substantive discussion. Instead of directly relating technology to existential dimensions of human life, it tries to describe what effect it has on human action and on the natural world. What Weber termed rationalization is the extension of control by means of ordering and categorizing forms of human action and by introducing means of surveillance to secure that actual activities would follow these categories. A similar argument could be developed for ordering, categorization, and action on the natural world.

The term "rationalization" clearly connotes a progressive aspect; I prefer to speak of formalization instead, to avoid this connotation and to stress the means by which this extension of control was achieved. Formalization means the reduction of a complex social or natural phenomenon by decomposing it into elements which are described such as to make them generic (i.e., to be found in other complex phenomena as well). "Technologies"—in a broad sense, including social technologies such as bureaucracy—recompose these elements in such a way that both the elements themselves and their relations are well describable and controllable.

I call this view "classic modern" because it is a variation of the more general theme that modernity provides its own means, and does not have to draw on sources external to itself, to order the world. As a reflected position, it seems, this particular variant surfaces only as a response to the criticism of technology and the appropriations of technology as revelation early in the twentieth century. Weber, a sober and usually hesitant analyst, saw this as the inevitable future, and even talked about "progress," but also lamented losses.[4] In the rationalist positivism of the Vienna Circle, in the architecture of Neue Sachlichkeit/International Style, in the debates on social and economic planning, and elsewhere, such classic modernism advanced to a strongly propagated normative position, a twentieth-century version of the ideal of progress, re-emerging from a first round of severe criticism. Owing to the fervor with which it was proposed during the interwar period, classic modernism even appeared to overcome its inability to thematize the alleged "irrationality" of its pure rationality. It was claimed that rationalization as desubstantivization really meant not the loss of substantive aspects of human existence but the overcoming of such tradition-bound obstacles to a well-ordered society.

However, such a sense of victory for modernism was premature. It kept being challenged both on grounds of its lack of any answer to the substantive question behind rationalities and on grounds of its incapability to deal with the experience of technology. The post–World War II progress in increasing the reach of human action through technical means had repercussions for both the philosophical debate and the political debate. Arendt (1958) and Heidegger (1993) used the example of a space traveler's distant view of the earth to rethink the conditions of human knowledge and action. (See also Tester 1995.) As it turned out, their remarks can now be read as an element of the opening of the debate on postmodernism. More concretely, technological rationalities are being debated as a political topic in environmentalism.

And despite the stunning development of modern technology itself, modernist theorizing remained equally incapable of grasping the fascination with the experience of technology. In the institutionally consolidated disciplines of the social sciences since World War II, most research on technology has adopted a very rationalistic approach. It is not my ambition here to move far toward reconceptualizing the sociology of

4. For Weber, though, as was briefly mentioned earlier, it is characteristic that he saw exactly this rationalization as an existential feature of the modern human condition and thus provided anchoring points for later critical theorists as diverse as Adorno and Marcuse or Heidegger.

technology. One step into this direction, however, is to move beyond attempts to construct a direct and comprehensive relation between technology and humankind or society (such approaches prevailed at the time under study here) and to differentiate between different social structures of technology experience and different kinds of technology without abandoning the notion of historically assessing the "human condition" with regard to technology. Certain elements of such a project can, indeed, be extracted from debates that took place early in the twentieth century.

Among the writings that stressed the novelty of the technological experience and its revelatory character, sometimes the individual aspect was emphasized. The possibility of new experiences allowed deep insights into the human condition, and it broadened and deepened the recognition of the self. In other writings, collective redemption was the focus of interest, the collectivity often being substantively defined as the nation or the working class. Significantly, the former view prevails in aesthetic, psychological, and philosophical debates, the latter in socio-political texts.

And, typically, the automobile and the airplane were technical examples for a reasoning of the first kind, the factory for the second. Futurism as a movement occupies a peculiar double position full of tensions between individualism and collectivism, which have its doubtful intellectual solution in Italian fascism as a national orientation valuing individual self-realization. And the city—as well as, to some extent, war—has a similarly double position among the technical examples, being evoked both for the anonymous hectic of dense collective life and for the freedom of the individual from imposed social ties and norms.

At this point, let us return to the socio-political issue of intellectual appropriation of technology. So far we may be said to have reconstructed three main positions: individual redemption, classic modern desubstantivization, and collective redemption. They were all proposed against the background of full inclusion of all members of a society into modernity, often known as mass society. Thus, they all dealt somehow with the abdication of the liberal idea of the citizen and thematized, even if sometimes rather implicitly, the relationship of individual and society.

Individual redemptionism, unless it remained content with the enhancement of individual self-realization, tended to develop a "new man" theory according to which energetic, creative persons would emerge who would lead society into new directions with the help of new technical means. Collective redemptionism held that the new technologies would forge strong collectivities which together would realize humankind's strivings. Jünger, for instance, found national-socialist followers who praised the building of a workers' state in the Soviet Union. Classic modernism developed a sort of conveyor-belt concept according to which technical and organizational elites would detect means of rationalization that would be to the immediate benefit of the followers as well as recognized as such. These views on technology were part of a profound restructuring of the political debate around the time of World War I.

World War I as a Divide in Political and Intellectual History

In a number of respects, World War I can be analyzed as a crucial period for the reorientation of intellectual work as well as, obviously, for political struggle. Ernst Jünger turned his experience of having been thrown into the technological hell of World War I into a widely read memoir. The futurists, who had praised the coming war before 1914, "perceived, accepted and supported the immersion of individuals into the reign of the machine [during the war] as necessary and indispensable for the test of technology" (Nazzaro 1987: 77). As we have seen, some observers regarded the war as the first large-scale application of advanced technology and, thus, as the inauguration of a new relation between man and technology. At the same time, the experience of wartime social and economic organization was often interpreted as showing the possibility and superiority of collective arrangements of organized cooperation, at least inside nations. This superiority stemmed from two distinct elements. First, the planned coordination of the wartime economy appeared more efficient than the rules of the market. Second, the cooperation between employers and workers' unions, supported and enforced by the state, seemed to point a way out of antagonistic class struggle. The occurrence of the war itself, furthermore, was regarded as the outcome of unfettered and unregulated workings of liberal-capitalist rules. The profit-driven development of the economy and of technology would lead people, classes, and nations into disastrous and violent competition and antagonism.

For all these reasons, the experience of World War I—and, importantly, of the Russian Revolution, itself not least a product of the war—reshaped the debates on a needed transformation of liberal capitalism that had been going on since at least the 1890s. It meant the temporary disappearance of liberalism as the organizer and focus of discourse.

Generally, it strengthened the "collectivist" position in these debates (i.e. the view that the "autonomous" development of economy and technology had to be regulated and controlled in the name of some higher-order reason and by some collective actor). What this higher-order reason should be ("common weal," "solidarity," "fate of the nation," and "existential features of humanity," to name a few possibilities), and who and what the actors or the arrangements could be (the state, corporate arrangements, the working class, intellectuals, engineers and technocrats, . . .), differed widely with political standpoints and also, significantly, between nations.

Indeed, to obtain a synthetic view of interwar politico-intellectual developments, I propose the image of a divergence of discursive trajectories. Before the war, as mentioned above, positions were loosely clustered around a conception of political liberalism. Even though the concept was not enthusiastically embraced, a widespread consensus on its inevitability prevailed. This consensus included the rising revisionist wings of socialist parties and their intellectuals as well as many would-be supporters of the persistent *ancien regimes* of the German and Austro-Hungarian empires (Mayer 1985), a fact that is telling enough of its evidence. The war and its aftermath liberated the centrifugal elements in this discursive regime. The emerging intellectual diversity is, to some extent, reflected in the diversity of the political regimes that were constructed as responses to the decline of liberalism. Historically, Soviet socialism, German National Socialism, the Swedish "People's Home," the French Popular Front, and the American New Deal can all be considered as varieties of such responses. World War II and its aftermath eliminated some of these responses, most notably National Socialism and Italian (but not Spanish) fascism, but it also eliminated the possibility of socialist regimes in the West. Intellectually, it limited the variation of thinking about technology, too.

Appropriations of Technology and the Social Order: America as a Threat and as a Solution

One main line in the European debates on technology, as I have tried to demonstrate, stressed the high importance of technology for social developments, positively or negatively, and tended to favor a collective approach, controlling and directing technology with a view to avoiding the worst of the dangers or realizing all of the promises (depending on the position). Inasmuch as such reorganization was considered as a conscious and planned effort, specific groups of actors were proposed and offered themselves as guiding elites for such a process. Often such proposals emanated from engineers' circles and were—implicitly or explicitly—linked to ideas of technocracy—societal steering by a scientific-technical elite (Dierkes et al. 1990). More generally, they gave new meaning to the question of political agency in a "mass society," in which conceptions of the responsible individual appeared superseded.

In the context of continental Europe, where the centrality of the state was still relatively unshaken, such debate was almost inevitably focused on the state. The state appeared as a natural addressee and actor when technological regulation was at issue.[5] In much of the European tradition, even though a number of different understandings of the state and its relation to social groups can be identified, the state was seen not merely as an institution the rules of which happened to apply to all individuals but as the comprehensive representative of a unity that was much more than its constituent parts. The inclination toward a collectivist response was further enhanced by the fact that Europeans tended to see their societies as being exposed to technologies which they did not themselves produce, which came from the outside and forced a reaction on them.

The view that the contested technologies mainly came from the United States, and that the United States was somehow essentially different from Europe, provided something of a common substantive background to the European debates on technology. In a review of interwar attitudes toward technologies, Overy (1990: 74) goes so far as to claim that "the fears we have described were not American fears." As far back as the 1920s, Antonio Gramsci (cited in Maier 1970: 27) wrote: "The European reaction to Americanism . . . must be examined attentively. Analysis of it will provide more than one element necessary for understanding the present situation of a series of states of the old continent and the political events of the post-war period." A look at "America" as a topic in these debates will help us to understand some of the European attitudes; at the same time, observations on the American way of dealing with technol-

5. See Herf 1984 and Breuer 1992: 104–106) for Germany. The technology debate merged with a broader discussion on the necessity and feasibility of economic planning to replace or complement market mechanisms. A look at research by Matthias von Bergen (1995) has given me an impression of the hegemony of the planning idea during the 1930s as well as of the variety of conceptions of planning.

ogy will provide some help in grasping the slowly emerging dominance of the classic modern view.

In all European countries, American technological developments, and especially the production technologies developed by Frederick Winslow Taylor and Henry Ford, were regarded as highly superior to European ones. Often they seen were not only as a limited technical example but as some sort of a model for economic and social reorganization. Thus, they provided opportunities for discussing nationally specific paths to reorganization. In Germany, for instance, Friedrich von Gottl-Ottlilienfeld (1924) praised Henry Ford's "white socialism" as an ethically sound application of "technical reason" to the betterment of society.

In the context of an analysis of appropriations of technology, Taylorism and Fordism are significant in several respects. First, they mark a clear difference between national situations according to the state of the industrial economy, thus allowing us to contextualize ideas about a national way to technology. (See chapter 5 above.) In the realm of industrial production, many European statements agree on both a general appreciation of US technical advances and the need for national adaptation of those concepts. Lysis (1917) demanded the national organization of the French productive system around the "idea of a national technology."[6] In 1931, an observer of changes in French car production concluded that Louis Renault "does not create an imitation of America" but "adapts the machines to the French needs and to our system" (Boulogne 1931, cited in Fridenson 1972: 55). In Italy, Boccioni talked about "the need for us to Americanize" in order to enter the realm of modernity (Gentile 1988: 114). Second, Taylor's "scientific work organization" and Ford's production concept, as means of formalization and thus of control and order through technology, introduced new questions about the relations between social classes, or between elites and followers, to the societies under study here. Early attempts to introduce the Taylor system were often rejected by the workers and countered by strikes— partly successfully, such as with Renault's first attempt (Le Chatelier). However, the view that these techniques could put an end to the zero-sum game of class dispute and form the beginning of a post-bourgeois age in which scientific rationality would be used to the benefit of all spread very fast (Maier 1970: 43). If a debate was led on such terms, transcendental aspects could be deemphasized. One could work in a modernist big factory without becoming Jünger's worker and without

6. "Lysis" was a pseudonym of Eugene Letailleur.

creating his worker's state. Labor unions and employers gradually moved to such positions in Europe during the interwar period and adopted them almost completely from the 1950s on. One aspect of the technology question had thus been transformed into a bargaining situation between major social actors in the context of society-wide feasible economic strategies.[7]

The factory, however, provided only one of the topics of the technology debate, though an important one. An equally important topic was what might be called the technization of the everyday world. The reference to the United States was as widespread here as in the case of Taylorism and Fordism. European or Europeanized portraits of America were littered with descriptions of technical wonders, and often the point of such descriptions was the emergence of an essentially different, alien form of life. (See chapter 7 above.)

Under the title *America, the Menace: Scenes from the Life of the Future* Georges Duhamel characterized the United States as "another civilization that is predominantly mechanical" (cited in Mathy 1993: 55). Paul Claudel, a writer who became France's ambassador to the United States, always remained nostalgic about Europe but was also impressed by modern techniques—sometimes in ways reminiscent of futurism. In 1928 Claudel described the United States as "a dynamo inserted between the two poles and the two ends of the continent."[8] And in the same text we find the following account of technical experiences: "Movement is everywhere and cities are the power-plants which supply it. . . . The car and the cinema are similar in principle. With one, motionless nature is transformed through our own movement in some kind of colorful wind. With the other, we remain seated and inexhaustible masses of ghosts charge into us." Beyond the experience, Claudel writes, these instruments transform the human condition essentially: "We are no longer subjected to circumstances, we dominate a text, we walk in the cosmos."

Most European statements on the United States touch on fundamental questions of human existence. Europeans are impressed or shocked by the way Americans handle those issues—sometimes they observe a technical, mechanical distortion of the questions; sometimes they seem to suggest that Americans have abandoned them entirely. In the context of my argument, this difference can be read in other terms. First, it

7. See Telo 1988: 27 on the "nationalization of social democracy" during the planning debates of the 1930s.

8. All these quotations are from *Conversations dans le Loir-et-Cher*, quoted after Mathy (1993: 87f.). For similar German quotations see Berg 1963 and Schwan 1986.

seems obvious that the enablements provided by technology are much less reluctantly embraced in the United States. There are fewer principled objections, and the result is the technical advance as observed by Europeans. Second, there is no evidence that the American producers and users of these technologies associate them less with existential feelings and experiences than Europeans (see chapter 4 above), as some of the latter sometimes suspected. Third, however, despite a debate that was not much less extended than in Europe, Americans were somewhat less obsessed with the need to link the common elaboration of a standpoint on technology to their collective destiny. One might say that transcendental matters, at least some of them, were privatized. They were open and subject to debate without an enforced need to come to a common conclusion and act accordingly.

Such a situation is not equal to embracing a classic modern view. Indeed, as Jamison shows in chapter 4 above, a variety of foundational arguments circulated in the American debate. However, it leaves the resort to a classic modern emphasis on functionality and "comfort," as to any other view, open to individual and group decision. Thus, deliberation on matters of principle is largely removed from collective debate, and impediments to technical development are reduced in actual practice. The resulting discursive situation may be characterized as the dissolution of the technology question into various aspects and elements that could hardly be reassembled to meet the demand for one principled answer.

Some technologies were indeed to be used collectively, as in work and war, but they and their effects came to be considered to touch only a part of human existence. They were regarded as functionally necessary and as existentially not too significant. At the same time, everybody was free to endow technological experience with higher meaning. One might embrace it like a new religion or reject it as dangerous to the human fate and mission; but this view was then considered a private move, without collective repercussions, since the collectivity, the polity did not—and did not need to—take any stand. This was a compromise that was temporarily accepted in the United States and, after World War II, in Europe. It formed the background to the immense diffusion and further development of technologies during the past four decades. It avoided the resort to strong solutions, such as a Heideggerian or a Jüngerian one, but it did so at the cost of repressing salient issues that were bound to re-emerge. The technology debates of the past two decades, an analysis of which is beyond the scope of this volume, mark such a re-emergence.

Technology and Contingency in a Long-Term Perspective

Even without analyzing the discursive structure of those present debates, we can draw one conceptual conclusion from the technology debate that took place earlier in this century and from its historical outcome. The historical outcome—by which I mean the relative dominance and political persuasiveness of the classic modern view of technology—lends itself to a very clear-cut conception of technology and discourses of technology nowadays. Technologies are assessed according to the enablements they provide in terms of the extension of reach and the enhancement of control in human action. And technology discourses are distinguished according to their acceptance of this conceptualization. On the one hand, technological development is seen, and accepted, as both a means and an effect of the functional differentiation of realms of human action. On the other hand, a moralist fundamentalism keeps raising issues that cannot be handled on modernist terms. Such theorizing has to, and will, be marginalized, although reflective observers recognize that it will not disappear (van den Daele 1992).

Such counterposition of incompatible theories does nothing but reproduce historical appropriations of technology, and, in view of the way the alternative is constructed, it can hardly escape an endorsement of the former perspective. As such, it appears to socially validate a historical outcome; however, it is very far from a theoretically sound conceptualization. To put it more strongly, it misses the most important point, which its construction made disappear. It does not recognize that the classic modern view, which promises to desubstantivize matters and to handle technology on purely rationalist and instrumental terms, itself takes a substantive stand—namely one that excludes considerations of the "human condition" (the conditions and meaning of human and social life) from the debate about technology. For an adequate conceptualization of technology, it is not sufficient merely to note that such considerations might always resurface as explicit topics of debate. Their existence and re-emergence touches on the core of "technology" itself. The uses and effects of technology have to do with the transformation of the conditions of human action (on whatever terms: toward predictability or creativity, toward autonomy or control) and are not just means to given ends. But if this is so, then technology needs to be a key topic of social theory, of historical analysis, and of politics; it cannot be contained in a functional frame around which socio-political debate moves without entering the core.

To put the matter in other terms, there seems to be broad agreement in modernist social science that the early-twentieth-century debate on technology was theoretically not very fruitful. Such a view would emphasize, not entirely without justification, that the "machines" of that period ultimately did not transform society itself into a machine, either in the Jüngerian sense of a new age or in the terms of critical theory as the end of the individual and the ultimate decline of political action. And once that is said, one may conclude that more recent technologies, such as "artificial intelligence" and genetic engineering, are not likely to have such millenarian or apocalyptic effects either. However, against such a standpoint, and with the help of a return to the earlier debates as pursued in this volume, it must be underlined that technologies have significantly transformed the conditions of human action, and may do so even more in the future. A modernist social science that excludes basic theoretical and normative considerations from view will not be able to understand these transformations or the emergence of technologies and their effect. "The most fundamental aspect of our culture," as Latour (1991: 35) puts it strongly, is that "we live in societies whose social link is laboratory-made objects."

Bibliography

Abramowski, Günter. 1966. *Das Geschichtsbild Max Webers. Universalgeschichte am Leitfaden des okzidentalen Rationalisierungsprozesses.* Ernst Klett.

Aglietta, Michel. 1979. *A Theory of Capitalist Regulation: The US Experience.* NLB.

Åkerman, Brita, ed. 1983a. *Den okända vardagen. Om arbetet i hemmen.* Akademilitteratur.

Åkerman, Brita. 1983b. *Vi kan vi behövs! Kvinnorna går samman i egna föreningar.* Akademilitteratur.

Alchon, Guy. 1985. *The Invisible Hand of Planning: Capitalism, Social Science, and the State in the 1920s.* Princeton University Press.

Altena, B. 1993. "Zu den Wirkungsbedingungen des niederländischen Sozialismus 1870–1914." In *Freiheitsstreben, Demokratie, Emanzipation. Aufsätze zur politischen Kultur in Deutschland und den Niederlanden,* ed. H. Lademacher and W. Mühlhausen. Lit.

Althin, Torsten. 1958. *Axel F. Enström. En minnesbok.* Ingeniörsvetenskapsakademien.

Åmark, Klas. 1990. Anteckningar om Max Weber, politik och professioner. Manuscript, SCASSS, Uppsala.

Arendt, Hannah. 1958. *The Human Condition.* University of Chicago Press.

Aristotle. 1981. *The Politics.* Penguin.

Arrow, Kenneth, et al. 1995. "Economic growth, carrying capacity, and the environment." *Science* 268: 520–521.

Åström, Lissie. 1985. "Husmodern möter folkhemmet." In *Modärna tider,* ed. J. Frykman and O. Löfgren. Liber Förlag.

Baark, Erik, and Andrew Jamison, eds. 1986. *Technological Development in China, India and Japan: Cross-cultural Perspectives.* St. Martin's.

Banta, Martha. 1993. *Taylored Lives: Narrative Productions in the Age of Taylor, Veblen, and Ford.* University of Chicago Press.

Barnouw, Dagmar. 1988. *Weimar Intellectuals and the Threat of Modernity.* Indiana University Press.

Baum, Richard. 1977. "Diabolus ex machina: Technological development and change in Chinese industry." In *Technology and Communist Culture*, ed. F. Fleron. Praeger.

Bauman, Zygmunt. 1989. *Modernity and the Holocaust.* Cornell University Press.

Bauman, Zygmunt. 1991. *Modernity and Ambivalence.* Polity Press.

Beetham, David. 1985. *Max Weber and the Theory of Modern Politics.* Polity Press.

Bell, Daniel. 1971. "Technology and politics." *Survey* 17: 9–13.

Bell, Daniel. 1993. "The breakdown of time, space and society." *Frontier International* 3: 9–13.

Beniger, James R. 1986. *The Control Revolution: Technological and Economic Origins of the Information Society.* Harvard University Press.

Berg, Maxine. 1980. *The Machinery Question and the Making of Political Economy.* Cambridge University Press.

Berg, Peter. 1963. *Deutschland und Amerika 1918–1929. Über das deutsche Amerikabild der zwanziger Jahre.* Matthiesen.

Berman, Marshall. 1982. *All That Is Solid Melts into Air: The Experience of Modernity.* Simon & Schuster.

Bernal, John. 1939. *The Social Function of Science.* Macmillan.

Berner, Boel. 1981. *Teknikens värld. Teknisk förändring och ingenjörsarbete i svensk industri.* Arkiv.

Bijker, Wiebe E. 1995a. "Sociohistorical technology studies." In *Handbook of Science and Technology Studies*, ed. S. Jasanoff et al. Sage.

Bijker, Wiebe E. 1995b. *Of Bicycles, Bakelite, and Bulbs: Toward a Theory of Sociotechnical Change.* MIT Press.

Björck, Henrik. 1992. *Teknikens art och teknikernas grad. Föreställningar om teknik, vetenskap och kultur speglade i debatterna kring en teknisk doktorsgrad, 1900–1927.* Royal Institute of Technology, Stockholm.

Blake, Casey. 1990. *Beloved Community: The Cultural Criticism of Randolph Bourne, Van Wyck Brooks, Waldo Frank and Lewis Mumford.* University of North Carolina Press.

Bloemen, E. S. A. 1988. *Scientific Management in Nederland.* NEHA.

Bok, Sissela. 1987. *Alva*. Bonniers.

Bollenbeck, Georg. 1994. *Bildung und Kultur. Glanz und Elend eines deutschen Deutungsmusters*. Insel.

Boltanski, Luc. 1987. *The Making of a Class: Cadres in French Society*. Cambridge University Press.

Boorstin, Daniel. 1987. *Hidden History: Exploring Our Secret Past*. Harper & Row.

Boschloo, T. J. 1989. *De productiemaatschappij*. Verloren.

Bose, Christine. 1979. "Technology and changes in the division of labour in the American home." *Women's Studies International Quarterly* 2: 295–304.

Bourdieu, Pierre. 1980. *The Logic of Practice*. Stanford University Press (1990).

Bourdieu, Pierre. 1991. *Language and Symbolic Power*. Harvard University Press.

Brautigam, J. 1956. *Langs de havens en op de schepen. Herinneringen*. Arbeiderspers.

Breuer, Stefan. 1992. *Die Gesellschaft des Verschwindens. Von der Selbstzerstörung der technischen Zivilisation*. Julius.

Bridenthal, Renate, Athina Grossman, and Marion Kaplan, eds. 1984. *When Biology Became Destiny*. Monthly Review Press.

Broeze, F. 1991. "Militancy and pragmatism: An international perspective on maritime labor." *International Review of Social History* 36: 165–200.

Brown, Marjorie M. 1985. *Philosophical Studies of Home Economics in the United States*. Michigan State University Press.

Bulletin, The. 1972. "History of the International Federation for Home Economics." No. 2–3. Secrétariat Général, Federation Internationale Pour L'Economie Familial, Paris.

Bussemer, H. U., S. Meyer, Barbara Orland, and E. Schulze. 1988: "Zur technischen Entwicklung von Haushaltgeräten und deren Auswirkung auf de Familie." In *Arbeitsplatz Haushalt*, ed. G. Tornieporth. Dietrich Reimer.

Camp, Richard L. 1969. *The Papal Ideology of Social Reform*. Brill.

Carlson, W. Bernard. 1992. "Artifacts and frames of meaning: Thomas A. Edison, his managers, and the cultural construction of motion pictures." In *Shaping Technology/Building Society*, ed. W. Bijker and J. Law. MIT Press.

Carlsson, Christina. 1986. *Kvinnosyn och kvinnopolitik. En studie av svensk socialdemokrati 1890–1910*. Archiv.

Castoriadis, Cornelius. 1990. *Le monde morcel. Les carrefours du labyrinthe III*. Seuil.

Chandler, Alfred D. 1977. *The Visible Hand: The Managerial Revolution in American Business*. Harvard University Press.

Chandler, Alfred. 1990. *Scale and Scope. The Dynamics of Industrial Capitalism.* Harvard University Press.

Childers, Thomas. 1990. "The social language of politics in Germany: The sociology of political discourse in the Weimar Republic." *American Historical Review* 95: 331–358.

Childs, Marquis W. 1936. *Sweden: The Middle Way.* Yale University Press.

Central Intelligence Agency, US. 1949. *An Estimate of Swedish Capabilities in Science.*

Clarke, Robert. 1973. *Ellen Swallow: The Woman Who Founded Ecology.* Follet.

Cocheret, Charles A. 1933. *Het elevator-bedrijf in de Rotterdamsche haven, 1908–1933.* Nijgh en Van Ditmar.

Cockburn, Cynthia, and Ruza Fürst-Dilic, eds. 1994. *Bringing Technology Home: Gender and Technology in a Changing Europe.* Open University Press.

Cockburn, Cynthia, and Susan Ormrod. 1993. *Gender and Technology in the Making.* Sage.

Colbjørnsen, Ole. 1934. "Planøkonomi i praksis. Den første og den annen treårsplan." *Det 20de århundre*, no. 9.

Collier, Peter, and David Horowitz. 1987. *The Fords—An American Epic.* Summit Books.

Collins, Randall. 1986. *Weberian Sociological Theory.* Cambridge University Press.

Collotti, Enzo. 1990. "Nationalism, anti-semitism, socialism and political Catholicism as expressions of mass politics in the twentieth century." In *Fin de Siecle and Its Legacy*, ed. M. Teich and R. Porter. Cambridge University Press.

Coudenhove-Kalergi, Richard Nikolaus. 1922. *Apologie der Technik.* Neue Geist Verlag.

Cowan, Ruth Schwartz. 1983. *More Work for Mother: The Ironies of Housework Technology from the Open Hearth to the Microwave.* Basic Books.

Crol, W. A. H. 1947. *Een tak van de familie Van Stolk. Honderd jaar in de graanhandel, 1847–1947.* Van Waesberge, Hoogewerff en Richards.

Cutcliffe, Stephen, and Robert Post, eds. 1989. *In Context: History and the History of Technology.* Lehigh University Press.

Dann, Otto. 1993. *Nation und Nationalismus in Deutschland 1770–1990.* Beck.

Daston, Lorraine. 1995. "The moral economy of science." *Osiris* 10: 3–24.

De Felice, Renzo, ed. 1988. *Futurismo, cultura e politica.* Fondazione Giovanni Agnelli.

De Geer, Hans. 1978. *Rationaliseringsrörelsen i Sverige. Effektivitetsidéer och socialt ansvar under mellankrigstiden.* Studieförbundet näringsliv och samhälle, Stockholm.

Dessauer, Friedrich. 1908. *Technische Kultur.* Kösel.

Dessauer, Friedrich. 1958. *Streit um die Technik.* Josef Knecht.

Dewey, John. 1939. *Freedom and Culture.* Putnam.

Dewey, John. 1929. "The quest for certainty." In *The Later Works. 1925–1953,* volume 4. Southern Illinois University Press (1984).

Dienel, Hans-Liudger. 1995. *Ingenieure zwischen Hochschule und Industrie. Kältetechnik in Deutschland und Amerika, 1870–1930.* Vandenhoeck & Ruprecht.

Dierkes, Meinolf, Andreas Knie, and Peter Wagner. 1988. "Die Diskussion über das Verhältnins von Technik und Politik in der Weimarer Republik." *Leviathan* 16: 1–22.

Dierkes, Meinolf, Andreas Knie, and Peter Wagner. 1990. "Engineers, intellectuals and the state." *Industrial Crisis Quarterly* 4 : 155–174.

Dierkes, Meinolf, Ute Hoffmann, and Lutz Marz. 1996. *Visions of Technology: Social and Institutional Factors Shaping the Development of New Technologies.* Campus and St. Martin's Press.

Dietz, Burkhard, Michael Fessner, and Helmut Maier, eds. 1996. *Technische Intelligenz und "Kulturfaktor Technik." Kulturvorstellungen von Technikern und Ingenieuren zwischen Kaiserreich und früher Bundesrepublik Deutschland.* Waxmann.

Di Maggio, Paul, and W. W. Powell. 1991. "Introduction." In *The New Institutionalism in Organizational Analysis,* ed. W. Powell and P. Di Maggio. University of Chicago Press.

Dreyfus, Hubert. 1995. "Heidegger on gaining a free relation to technology." In *Technology and the Politics of Knowledge,* ed. A. Feenberg and A. Hannay. Indiana University Press.

Dunlavy, Colleen A. 1994. *Politics and Industrialization: Early Railroads in the United States and Prussia.* Princeton University Press.

Durkheim, Emile. 1959. *Socialism and Saint-Simon.* Antioch.

Edwards, Richard. 1979. *Contested Terrain: The Transformation of the Workplace in the Twentieth Century.* Prentice-Hall.

Ehrenreich, Barbara, and Deidre English. 1979. *For Her Own Good.* Doubleday Anchor.

Eldorado: Homosexuelle Frauen und Männer in Berlin 1850–1950. Geschichte, Alltag und Kultur. 1984. Frölich & Kaufmann.

Ellul, Jacques. 1964. *The Technological Society.* Knopf.

Elzinga, Aant. 1993. "Universities, research, and the transformation of the state in Sweden." In *The European and American University*, ed. S. Rothblatt and B. Wittrock. Cambridge University Press.

Ewald, François. 1986. *L'Etat-providence*. Grasset.

Everwijn, J. C. A. 1912. *Beschrijving van handel en nijverheid in Nederland*. Belinfante.

Eyerman, Ron. 1985. "Rationalizing intellectuals: Sweden in the 1930s and 1940s." *Theory and Society* 14: 777–808.

Eyerman, Ron. 1992. "Intellectuals and the state: A framework for analysis, with special reference to the United States and Sweden." In *Vanguards of Modernity*, ed. N. Kauppi and P. Sulkunen. Research Unit for Contemporary Culture, Jyväskylä.

Eyerman, Ron. 1994. *Between Culture and Politics: Intellectuals in Modern Society*. Polity Press.

Eyerman, Ron and Andrew Jamison. 1989. "Environmental knowledge as an organizational weapon: The case of Greenpeace." *Social Science Information* 28: 1: 99–119.

Eyerman, Ron, and Andrew Jamison. 1991. *Social Movements: A Cognitive Approach*. Polity Press.

Ezrahi, Yaron, Everett Mendelsohn, and Howard P. Segal, eds. 1995. *Technology, Pessimism and Postmodernism*. Kluwer.

Feder, Gottfried. 1934. *Hitler's Official Programme and Its Fundamental Ideas*. Allen & Unwin.

Feenberg, Andrew. 1991. *Critical Theory of Technology*. Oxford University Press.

Ford, Henry. 1926. *Today and Tomorrow*. Productivity Press (1988).

Forman, Paul. 1971. "Weimar culture, causality, and quantum theory." *Historical Studies in the Physical Sciences* 3: 1–116.

Foucault, Michel. 1966. *The Order of Things: An Archaeology of the Human Sciences*. Random House (1971).

Fox, Bonnie J. 1990. "Selling the mechanized household: 70 years of ads in *Ladies Home Journal*." *Gender and Society* 4, no. 1: 25–40.

Fox, R., and Lears, T. eds. 1983. *The Culture of Consumption: Critical Essays in American History 1880–1980*. Pantheon.

Fox, Stephen. 1985. *The American Conservation Movement: John Muir and His Legacy*. University of Wisconsin Press.

Fraser, Steve, and Gary Gerstle, eds. 1989. *The Rise and Fall of the New Deal Order, 1930–1980*. Princeton University Press.

Freeman, Christopher, ed. 1984. *Long Waves in the World Economy*. Frances Pinter.

Fridenson, Patrick. 1972. "L'idéologie des grands constructeurs dans l'entre-deux-guerres." *Le mouvement social*, no. 81 (October–December): 51–68.

Fridenson, Patrick. 1978. "The coming of the assembly line to Europe." In *The Dynamics of Science and Technology*, ed. W. Krohn et al. Sociology of the Sciences Yearbook. Reidel.

Fromm, Erich. 1941. *Escape from Freedom*. Holt, Rinehart, and Winston.

Galbraith, John K. 1967. *The New Industrial State*. New American Library.

Gårdlund, Torsten. 1942. *Industrialismens samhälle*. Tiden.

Gentile, Emilio. 1988. "Il futurismo e la politica. Dal nazionalismo modernista al fascismo 1909–1920." In *Futurismo, cultura e politica*, ed. R. De Felice. Fondazione Giovanni Agnelli.

Gerth, H. H., and C. Wright Mills, eds. 1946. *From Max Weber: Essays in Sociology*. Oxford University Press.

Giddens, Anthony. 1984. *The Constitution of Society*. Polity Press.

Giddens, Anthony. 1990. *The Consequences of Modernity*. Polity Press.

Gilbreth, Lillian Moller. 1929. *The Home-Maker and Her Job*. Appleton.

Gispen, Kees. 1989. *New Profession, Old Order: Engineers and German Society, 1815–1914*. Cambridge University Press.

Gramsci, Antonio. 1971. *Selections from the Prison Notebooks*. Lawrence & Wishart.

Graswinckel, D. P. M., and Ott, L. 1973. *100 jaar "in granen." Handel en wandel van het Comité van graanhandelaren*. Comité van Graanhandelaren.

Green, Martin. 1974. *The von Richthofen Sisters: the Triumphant and the Tragic Modes of Love—Else and Frieda von Richthofen, Otto Gross, Max Weber, and D. H. Lawrence in the Years 1870–1970*. Weidenfeld and Nicholsen.

Grenstad, Gunnar, and Per Selle. 1995. "Cultural theory and New institutionalism." *Journal of Theoretical Politics* 7: 5–27.

Gribling, J. P. 1975. "Katholieke sociale actie in Europa en in Nederland." *Politiek perspectief* 4: 3–38.

Gross, David. 1992. *The Past in Ruins: Tradition and the Critique of Modernity*. University of Massachusetts Press.

Grüner, Gustav. 1967. *Die Entwicklung der höheren technischen Fachschulen im deutschen Sprachgebiet*. Westermann.

Grüttner, M. 1984. *Arbeitswelt an der Wasserkante. Sozialgeschichte der Hamburger Hafenarbeiter 1886–1914*. Vanderhoeck und Ruprecht.

Habbakuk, H. J. 1967. *American and British Technology in the Nineteenth Century: The Search for Labour-Saving Inventions.* Cambridge University Press.

Haber, Samuel. 1964. *Efficiency and Uplift: Scientific Management in the Progressive Era 1890–1920.* University of Chicago Press.

Habermas, Jürgen. 1981. *Theorie des kommunikativen Handels.* Suhrkamp.

Habermas, Jürgen. 1987. *The Philosophical Discourse of Modernity.* MIT Press.

Hagberg, Jan-Erik. 1986. *Tekniken i kvinnornas händer. Hushållsarbete och hushållsteknik under tjugo- och trettiotalen.* Liber Förlag.

Halvorsen, Tor. 1991. Tekniske yrker og organisasjon. AHS 1991-4. University of Bergen.

Halvorsen, Tor. 1993. Profesjonalisering og profesjonspolitikk. Den sosiale konstruksjon av tekniske yrker. Doctoral dissertation, Department of Adminstration and Organizational Science, University of Bergen.

Haraldsson, Désirée. 1987. *Skydda vår natur! Svenska Naturskyddsföreningens framväxt och tidiga utveckling.* Lund University Press.

Hård, Mikael. 1989. "History of technology in Sweden—A field with a future!?" *Polhem. Tidskrift för teknikhistoria* 7: 164–82.

Hård, Mikael. 1990. "Teknik-och-kultur-diskursen. Ett utsnitt ur den tyska moderniseringsdebatten, 1906–33." In *Teknik som diskurs. Moderniseringsdebatter i Tyskland och Sverige, 1905–35,* s. t. i. c., volume 3, ed. M. Hård and C. Mithander. Department of Theory of Science, Gothenburg University.

Hård, Mikael. 1993. "Beyond harmony and consensus: A social conflict approach to technology." *Science, Technology and Human Values* 18: 408–432.

Hård, Mikael. 1994a. "Technology as practice: Local and global closure processes in diesel-engine design." *Social Studies of Science* 24: 549–585.

Hård, Mikael. 1994b. *Machines Are Frozen Spirit: The Scientification of Refrigeration and Brewing in the 19th Century—A Weberian Interpretation.* Campus and Westview.

Hård, Mikael, and Conny Mithander, eds. 1990. *Teknik som diskurs. Moderniseringsdebatter i Tyskland och Sverige, 1905–35,* s. t. i. c., volume 3. Department of Theory of Science, Gothenburg University.

Hardach, Gerd. 1977. *The First World War. 1914–1918.* Pelican.

Harris, Abram L. 1958. *Economics and Social Reform.* Harper & Brothers.

Harvey, David. 1989. *The Condition of Postmodernity.* Blackwell.

Hayden, Dolores. 1981. *The Grand Domestic Revolution: A History of Feminist Designs for American Homes, Neighborhoods, and Cities.* MIT Press.

Hays, Samuel P. 1959. *Conservation and the Gospel of Efficiency: The Progressive Conservation Movement 1890–1920.* Harvard University Press.

Hegel, Georg W. F. 1991. *The Philosophy of History.* Prometheus Books.

Heidegger, Martin. 1959. *Introduction to Metaphysics.* Yale University Press.

Heidegger, Martin 1982. *Die Technik und die Kehre.* Neske.

Heidegger, Martin. 1993. "'Only a god can save us': *Der Spiegel*'s interview with Martin Heidegger 1966." In *The Heidegger Controversy,* ed. R. Wolin. MIT Press.

Helldén, Arne. 1986. *Maskinerna och lyckan. Ur industrisamhällets idéhistoria.* Ordfront.

Heller, Agnes, and Ferenc Fehér. 1988. "On being satisfied in a dissatisfied society II." In *The Postmodern Political Condition.* Polity Press.

Hellige, Hans Dieter. 1990. "Walther Rathenau: ein Kritiker der Moderne als Organisator des Kapitalismus." In *Ein Mann vieler Eigenschaften. Walther Rathenau und die Kultur der Moderne,* ed. T. Hughes et al. Klaus Wagenbach.

Hennebicque, Alain. 1992. "Albert Thomas and the war industries." In *The French Home Front, 1914–1918,* ed. P. Fridenson. Berg.

Herf, Jeffrey. 1984. *Reactionary Modernism: Technology, Culture, and Politics in Weimar and the Third Reich.* Cambridge University Press.

Hermand, Jost, and Frank Trommler. 1988. *Die Kultur der Weimarer Republik.* Fischer.

Hickman, L. 1990. *John Dewey's Pragmatic Technology.* Indiana University Press.

Hirdman, Yvonne. 1983. "Den socialistiska hemmafrun." In *Vi kan vi behövs! Kvinnorna går samman i egna föreningar,* ed. B. Åkerman. Akademilitteratur.

Hirdman, Yvonne. 1990. *Att lägga livet till rätta—studier i svensk folkhemspolitik.* Carlssons.

Hitler, Adolf. 1925, 1927. *Mein Kampf.* Eher.

Hobsbawm, Eric, and Terence Ranger, eds. 1985. *The Invention of Tradition.* Cambridge University Press.

Hofstadter, R. 1955. *The Age of Reform.* Knopf.

Holm, Ingvar. 1965. *Harry Martinson. Myter Målningar Motiv.* Bonniers.

Homburg, Heidrun. 1978. "Anfänge des Taylorsystems in Deutschlands." *Geschichte und Gesellschaft* 4: 170–194.

Hortleder Gerd. 1973. *Ingenieure in der Industriegesellschaft. Zur Soziologie der Technik und naturwissenschaftlich-technischen Intelligenz im öffentlichen Dienst und in der Industrie.* Suhrkamp.

Hortleder, Gerd. 1974. *Das Gesellschaftsbild des Ingenieurs. Zum politischen Verhalten der technischen Intelligenz in Deutschland.* Suhrkamp.

Huber, Joseph. 1989. *Technikbilder. Weltanschauliche Weichenstellungen der Technologie- und Umweltpolitik.* Westdeutscher Verlag.

Hueting, E., et al., eds. 1983. *Naar groter eenheid. De geschiedenis van het Nederlands Verbond van Vakverenigingen. 1906–1981.* Van Gennep.

Hughes, H. Stuart. 1958. *Consciousness and Society: The Reorientation of European Social Thought 1890–1930.* Knopf.

Hughes, Thomas P. 1981. "Ideologie für Ingenieure." *Technikgeschichte* 48: 308–323.

Hughes, Thomas P. 1983. *Networks of Power: Electrification in Western Society, 1880–1930.* Johns Hopkins University Press.

Hughes, Thomas P. 1987. "The evolution of large technological systems." In *The Social Construction of Technological Systems,* ed. W. Bijker et al. MIT Press.

Hughes, Thomas P. 1989. *American Genesis: A Century of Invention and Technological Enthusiasm.* Viking Penguin.

Hughes, Thomas P. 1990. "Walther Rathenau: 'System builder.' " In *Ein Mann vieler Eigenschaften. Walther Rathenau und die Kultur der Moderne,* ed. T. Hughes et al. Klaus Wagenbach.

Hughes, Thomas P., and Agatha C. Hughes, eds. 1990. *Lewis Mumford: Public Intellectual.* Oxford University Press.

Hughes, Thomas P., et al., eds. 1990. *Ein Mann vieler Eigenschaften. Walther Rathenau und die Kultur der Moderne.* Klaus Wagenbach.

Hultgren, Inger. 1982. *Kvinnors organisation och samhällets beslutsprocess.* Almqvist & Wiksell.

Idun. 1920. "De elektriska kökshjälpmedelns seger" (p. 880).

Idun. 1925. "Elektriciteten som husjungfru och köksa" (p. 264).

Idun. 1930. "Några funderingar inför oktoberflyttningen" (p. 997).

Jackson, W. A. Douglas. 1990. *Gunnar Myrdal and America's Conscience: Social Engineering and Racial Liberalism, 1938–1987.* University of North Carolina Press.

Jameson, Frederic. 1991. *Postmodernism, or The Cultural Logic of Late Capitalism.* Verso.

Jamison, Andrew. 1982. *National Components of Scientific Knowledge.* Research Policy Institute.

Jamison, Andrew. 1989. "Technology's theorists: Conceptions of technological innovation in relation to science and technology policy.'' *Technology and Culture* 30: 505–533.

Jamison, Andrew. 1991. ''National styles in technology policy: comparing the Swedish and Danish state programmes in microelectronics/information technology.'' In *State Policies and Techno-Industrial Innovation*, ed. U. Hilpert. Routledge.

Jamison, Andrew, and Ron Eyerman. 1994. *Seeds of the Sixties*. University of California Press.

Jan. 1935. ''Hur modern ungdom bo.'' *Idun*, no. 26: 27.

Jansen, T. 1979. '''De wil der bazen regelt het werk.' Havenarbeiders rond. 1900 in Rotterdam en Amsterdam.'' In *Jaarboek voor de geschiedenis van socialisme en arbeidersbeweging in Nederland*. Sun.

Jasanoff, Sheila. 1995. *Science at the Bar*. Harvard University Press.

Jarausch, Konrad H. 1990. *The Unfree Professions: German Lawyers, Teachers, and Engineers. 1900–1950*. Oxford University Press.

Joas, Hans. 1995. *The Creativity of Action*. Polity.

Joerges, Bernward. 1989. ''Soziologie und Maschinerie. Vorschläge zu einer 'realistischen Techniksoziologie.' '' in *Technik als sozialerProzess*, ed. P. Weingart. Suhrkamp.

Jörberg, Lennart. 1970. *The Industrial Revolution in Scandinavia*. Fontana Economic History of Europe, volume 4. Fontana.

Jünger, Ernst. 1929. *Feuer und Blut. Ein kleiner Ausschnitt aus dem grossen Schlacht*. Frundsberg.

Jünger, Ernst. 1932. *Der Arbeiter: Herrschaft und Gestalt*. Hanseatische Verlagsanstalt.

Kaelbe, Hartmut. 1989. ''Was Prometheus most unbound in Europe? The labour force in Europe during the late XIXth and XXth centuries.'' *Journal of European Economic History* 18: 65–105.

Kaj. 1930. ''Vårt dagliga bröd.'' *Idun*: 1215.

Kasson, J. 1977. *Civilizing the Machine: Technology and Republican Values in America, 1776–1900*. Penguin.

Kautsky, Karl. 1919. *Die Sozialiserung und die Arbeiterräte*. Sozialistische Bücherei, volume 5. Ignaz Brand & Co.

Kempf, R. 1923. ''Staat, Wirtschaft und Haushalt.'' *Die Frau* 31: 325–332.

Kenngott, Eva-Maria. 1990. *Der Organisationskulturanzsatz. Ein mögliches Programm zur Konzeption von Entscheidungsverhalten in Organisationen?* FS II 90-103. Wissenschaftszentrum Berlin für Sozialforschung.

Kern, Stephen. 1983. *The Culture of Time and Space, 1880–1918.* Harvard University Press.

Kimball, Bruce A. 1992. *The "True Professional Ideal" in America: A History.* Blackwell.

Kjellén, Rudolf. 1908. *Ett program.* Geber.

Kobayashi, Kesaji, and Hidemasa Morikawa, eds. 1986. *Development of Managerial Enterprise.* University of Tokyo Press.

Kocka, Jürgen. 1988. "German history before Hitler: The debate about the German *Sonderweg.*" *Journal of Contemporary History* 23: 3–16.

Koenne, Werner. 1979. "On the relationship between philosophy and technology in the German-speaking countries." In *The History and Philosophy of Technology,* ed. G. Bugliarello and D. Doner. University of Illinois Press.

Koonz, Claudia. 1984. "The competition for a Women's *Lebensraum.* 1928–1934." In *When Biology Became Destiny,* ed. R. Bridenthal et al. Monthly Review Press.

Koonz, Claudia. 1986. *Mothers in the Fatherland: Women the Family and Nazi Politics.* Methuen.

Krause, Werner. 1960. Werner Sombarts Weg vom Kathedersozialismus zum Faschismus. Dissertation, Humboldt University.

Krutch, Joseph Wood. 1929. *The Modern Temper: A Study and a Confession.* Harcourt, Brace.

Kyle, Gunhild, ed. 1987. *Handbok i svensk kvinnohistoria.* Carlssons.

Kylhammar, Martin. 1985. *Maskin och idyll. Teknik och pastorala ideal hos Strindberg och Heidenstam.* Liber.

Kylhammar, Martin. 1990. *Den okände Sten Selander. En borgerlig intellektuell.* Akademeja.

Kylhammar, Martin. 1994. *Frejdiga framstegsmän och visionära världsmedborgare: Epokskiftet 20-tal–30-tal genom fem unga och Lubbe Nordström.* Akademeja.

Lambourne, Robert, Michael Shallis, and Michael Shortland. 1990. *Close Encounters? Science and Science Fiction.* Adam Hilger.

Landes, David S. 1969. *The Unbound Prometheus: Technological Change and Industrial Development in Western Europe from 1750 to the Present.* Cambridge University Press.

Lange, Helene. 1919. "Die deutschen Frauen und der Frauenweltbund." *Die Frau* 27: 341–343.

Langguth, Frauke. 1989. Zur Geschichte der Elektrifisierung der privaten Haushalte. Die Absatzpolitik der BEWAG gegenüber den privaten Haushalten im

Berlin während der Weimarer Republik. Master's thesis, Technical University, Berlin.

Lasch, Christopher. 1991. *The True and Only Heaven: Progress and Its Critics.* Norton.

Latour, Bruno. 1991. *Nous n'avons jamais étés modernes.* La Découverte.

Layton, Edwin T. 1971. *The Revolt of the Engineers: Social Responsibility and the American Engineering Profession.* Press of Case Western Reserve University.

Leach, W. 1993. *Land of Desire: Merchants, Power and the Rise of a New American Culture.* Pantheon.

Leander, Sigfrid. 1978. *Folkbildning och folkföreläsningar.* LT.

Lears, T. 1981. *No Place of Grace: Antimodernism and the Transformation of American Culture 1880–1920.* Pantheon.

Lehner, Franz. 1983. "The vanishing of spontaneity: Socio-economic conditions of the welfare state." *European Journal of Political Research* 11: 437–447.

Lenk, Hans. 1982. *Zur Sozialphilosophie der Technik.* Suhrkamp.

Levine, L. 1993. *The Unpredictable Past: Explorations in American Cultural History.* Oxford University Press.

Lie, Merete, and Knut H. Sørensen, eds. 1996. *Making Technology Our Own? Domesticating Technology into Everyday Life.* Scandinavian University Press.

Liedman, Sven-Eric. 1984. "Om ideologier." In *Om ideologi och ideologianalys,* ed. S. E. Liedman et al. Department of History of Science and Ideas, Gothenburg University.

Liedman, Sven-Eric. 1986. *Den synliga handen. Anders Berch och ekonomiämnena vid 1700-talets svenska universitet.* Arbetarkultur.

Lilliehöök, Eleonor. 1930. Interview signed "B. A." *Idun:* 720.

Lindblom, Paul. 1991. *Samtiden i ögat. En bok om Artur Lundkvist.* Tiden.

Lindholm, Margareta. 1990. *Talet om det kvinnliga. Studier i feministiskt tänkande i Sverige under 1930-talet.* Monograph 44, Department of Sociology, University of Gothenburg.

Lindroth, Sten. 1952. *Swedish Men of Science.* Almqvist & Wiksell.

Lintsen, Harry, et al., eds. 1992–1995. *Geschiedenis van de techniek in Nederland.* Walburg.

Löfgren, Orvar. 1991. "Att nationalisera moderniteten." In *Nationella identiteter i Norden—ett fullbordat projekt?* ed. A. Linde-Laursen and J. Nilsson. Nordic Council.

Löfgren, Orvar. 1993. "Nationella arenor." In *Försvenskningen av Sverige. Det nationellas förvandlingar,* ed. B. Ehn et al. Natur och Kultur.

Lovell, J. 1969. *Stevedores and Dockers: A Study of Trade Unionism in the Port of London.* Macmillan.

Löwith, Karl. 1932. *Karl Marx and Max Weber.* Allen & Unwin (1982).

Ludwig, Karl Heinz. 1974. *Technik und Ingeniure im Dritten Reich.* Droste.

Lundqvist, Sven. 1977. *Folkrörelserna i det svenska samhället 1850–1920.* Almqvist & Wiksell.

Lüders, Marie Elisabeth. 1921. "Hat die Hausfrau einen Beruf?" *Die Frau* 28: 129–136.

Lyotard, Jean-Francois. 1984. *The Postmodern Condition: A Report on Knowledge.* University of Minnesota Press.

Lysis (Eugène Letailleur). 1917. *Vers la démocratie nouvelle.* Payot.

Macintyre, Alasdair. 1984. *After Virtue: A Study in Moral Theory.* University of Notre Dame Press.

Mackay, Hughie, and Gareth Gillespie. 1992. "Extending the social shaping of technology approach: Ideology and appropriation." *Social Studies of Science* 22: 685–716.

MacKenzie, Donald, and Judy Wajcman, eds. 1985. *The Social Shaping of Technology: How the Refrigerator Got Its Hum.* Open University Press.

March, James, and John Olsen. 1989. *Rediscovering Institutions: The Organizational Basis of Politics.* Free Press.

McCormick, R. L. 1981. "The discovery that business corrupts politics: A reappraisal of the origins of progressivism." *American History Review* 86: 247–274.

McNeill, William H. 1983. *The Pursuit of Power: Technology, Armed Force, and Society since A.D. 1000.* Blackwell.

McSweeney, B. 1980. *Roman Catholicism: The Search for Relevance.* Oxford University Press.

Mader, Ursula. 1974. *Walther Rathenau als Funktionär des Finanzkapitals. Beiträge zu einer politischen Biographie 1887–1917.* Humboldt University.

Maier, Charles S. 1970. "Between Taylorism and technocracy: European ideologies and the visions of industrial productivity in the 1920s." *Journal of Contemporary History* 5, no. 2: 27–61.

Maier, Charles S. 1987. *In Search of Stability: Explorations in Historical Political Economy.* Cambridge University Press.

Mann, Thomas. 1924. *The Magic Mountain.* Vintage Books (1969).

Mannheim, Karl. 1936. *Ideology and Utopia: An Introduction to the Sociology of Knowledge.* Routledge and Kegan Paul .

Marcuse, Herbert. 1968. *Negations: Essays in Critical Theory.* Penguin.

Martinson, Harry. 1956. *Aniara.* Bonniers.

Martinson, Harry. 1960. *Vagnen.* Bonniers.

Marx, Leo. 1964. *The Machine in the Garden: Technology and the Pastoral Ideal in America.* Oxford University Press.

Marx, Leo. 1988. *The Pilot and the Passenger: Essays on Literature, Technology and Culture in the United States.* Oxford University Press.

Marz, Lutz. 1993. *Mensch—Maschine—Moderne.* FS II 93-107. Wissenschaftszentrum Berlin für Sozialforschung.

Masini, Ferruccio. 1988. "Futurismo e rivoluzione conservatrice in Germania." In *Futurismo, cultura e politica,* ed. R. De Felice. Fondazione Giovanni Agnelli.

Mathy, Jean-Philippe. 1993. *Extreme-Occident. French Intellectuals and America.* University of Chicago Press.

Matthaei, Julie A. 1982. *An Economic History of Women in America.* Schocken.

Mayer, Arno. 1985. *The Persistence of the Old Regime.* Princeton University Press.

Meikins, Peter. 1988. "'The revolt of the engineers' reconsidered." *Technology and Culture* 29: 219–246.

Mendelsohn, Everett. 1990. "Prophet of our discontent: Lewis Mumford confronts the bomb." In *Lewis Mumford: Public Intellectual,* ed. T. Hughes and A. Hughes. Oxford University Press.

Merkle, Judith A. 1980. *Management and Ideology: The Legacy of the International Scientific Management Movement.* University of California Press.

Meurling, Per. 1984. *Den goda tonens pingviner.* Nya tidskrifts AB Folketbild.

Miller, Donald. 1989. *Lewis Mumford: A Life.* Weidenfeld and Nicolson.

Mitcham, Carl. 1994. *Thinking Through Technology: The Path between Engineering and Philosophy.* University of Chicago Press.

Mithander, Conny. 1990. "Reaktionär modernism i Sverige." In *Teknik som diskurs. Moderniseringsdebatter i Tyskland och Sverige, 1905–35,* s. t. i. c., volume 3, ed. M. Hård and C. Mithander. Department of Theory of Science, Gothenburg University.

Mithander, Conny. 1991. "Implications of technology: The debate in its context." In *Intellectuals Reading Technology,* ed. C. Landström. Department of Theory of Science, Gothenburg University.

Mitzman, Arthur. 1971. *The Iron Cage: An Historical Interpretation of Max Weber.* Knopf.

Mol, H. 1920. "Uit het Rotterdamsche havenbedrijf." *Socialistische Gids* 545–549, 655–662.

Mol, H. 1980. *Memoires van een havenarbeider.* Sun.

Molella, Arthur P. 1990. "Mumford in historiographical context." In *Lewis Mumford: Public Intellectual,* ed. T. Hughes and A. Hughes. Oxford University Press.

Molin, Adrian. 1909. "Sverige." In *Svenska allmogehem,* ed. G. Carlsson. Fritze.

Mommsen, Wolfgang J. 1974. *Max Weber and German Politics, 1890–1920.* University of Chicago Press (1984).

Moore, Martha Trescott. 1983. "Lillian Moller Gilbreth and the founding of modern industrial engineering." In *Machina Ex Dea,* ed. J. Rothschild. Pergamon.

Müler, Hans. 1979. "Frauen und Faschismus." In *Frauen in der Geschichte,* ed. A. Kuhn and G. Schneider. Schwann.

Müller, Hans Peter, and Ulrich Troitzsch, eds. 1989. *Technologie zwischen Fortschritt und Tradition. Beiträge zum Internationalen Johann Beckmann-Symposium Göttingen.* Peter Lang.

Mumford, Lewis. 1922. *The Story of Utopias.* Boni and Liveright.

Mumford, Lewis. 1926. *The Golden Day.* Beacon (1957).

Mumford, Lewis. 1934. *Technics and Civilization.* Harcourt, Brace.

Mumford, Lewis. 1982. *Sketches from Life: The Autobiography of Lewis Mumford, The Early Years.* Dial.

Myrdal, Alva. 1935. Interview. *Idun,* no. 45: 19.

Nazzaro, G. Battista. 1987. *Futurismo e politica.* JN Editore.

Nelson, Richard, ed. 1993. *National Innovation Systems: A Comparative Analysis.* Oxford University Press.

Nerad, Maresi. 1987. "Gender stratification in higher education: The Department of Home Economics at the University of California, Berkeley, 1916–1962." *Women's Studies International Forum* 10: 157–164.

Newell, W. R. 1988. "Politics and progress in Heidegger's philosophy of history." In *Democratic Theory and Technological Society,* ed. R. Day et al. Sharpe.

Niethammer, Lutz. 1989. *Posthistoire.* Rowohlt.

Nijhof, E. 1990. "Innovaties en werkgelegenheid in de haven van Rotterdam. 1890–heden." In *Geschiedenis tussen eigen ervaring en wetenschap.* VUlgo.

Nilsson, Jan Olof. 1994. *Alva Myrdal. En virvel in den moderna strömmen.* Symposion.

Noble, David. 1977. *America by Design: Science, Technology, and the Rise of Corporate Capitalism.* Knopf.

Nolan, Mary. 1994. *Visions of Modernity: American Business and the Modernization of Germany.* Oxford University Press.

Nolin, B., ed. 1993. *Kulturradikalismen. Det moderna genombrottets andra fas.* Symposion.

Nove, Alex. 1989. *An Economic History of the USSR.* Penguin.

Nyberg, Anita. 1989. Tekniken—kvinnornas befriare? Department of Technology and Social Change, Linköping University.

Nybom, Thorsten. 1993. "The Swedish social democratic state in a tradition of peaceful revolution." In *Konflikt og samarbejde. Festskrift til Carl-Axel Gemzell,* ed. C. Due-Nielsen et al. Museum Tusculanum.

Nye, David. 1991. *Electrifying America: Social Meanings of a New Technology, 1800–1940.* MIT Press.

Nye, David. 1994. *American Technological Sublime.* MIT Press.

Offe, Claus, and Helmut Wiesenthal. 1980. "Two logics of collective action." *Political Power and Social Theory* 1: 67–115.

Orr, John. 1974. "German social theory and the hidden face of technology." *Archives européennes de sociologie* 15: 312–336.

Ortega y Gassett, José. 1958. *History as a System and Other Essays.* Norton.

Our Common Future. 1987. World Commission on Environment and Development and Oxford University Press.

Overy, Richard. 1990. "Heralds of modernity: Cars and planes from invention to necessity." In *Fin de Siecle and Its Legacy,* ed. M. Teich and R. Porter. Cambridge University Press.

Pearson, Geoffrey. 1979. "Resistance to the machine." In *Counter-movements in the Sciences,* ed. H. Nowotny and H. Rose. Reidel.

Pells, R. 1973. *Radical Visions and American Dreams: Culture and Social Thought in the Depression Years.* Harper and Row.

Perry, L. 1989. *Intellectual Life in America: A History.* University of Chicago Press.

Peukert, Detlev. 1987. *Die Weimarer Republik. Krisenjahre der Klassischen Moderne.* Suhrkamp.

Peukert, Detlev. 1989. *Max Webers Diagnose der Moderne.* Vandenhoeck & Ruprecht.

Piore, Michael J., and Charles Sabel. 1984. *The Second Industrial Divide: Possibilities for Prosperity*. Basic Books.

Pippin, Robert. 1995. "On the notion of technology as ideology: Prospects." In *Technology, Pessimism and Postmodernism*, ed. Y. Ezrahi et al. 1995. Kluwer.

Pollard, Sidney. 1981. *Peaceful Conquest: The Industrialization of Europe, 1760–1970*. Oxford University Press.

Prinz, Michael, and Rainer Zitelmann. 1991. *Nationalsozialismus und Modernisierung*. Wissenschaftliche Buchgesellschaft.

Pursell, Carroll. 1979. "Government and technology in the Great Depression." *Technology and Culture* 20: 162–174.

Qvarsell, Roger, et al., eds. 1986. *I framtidens tjänst. Ur folkhemmets idéhistoria*. Gidlunds.

Rabinbach, Anson. 1996. "Social knowledge, fatigue, and the politics of industrial accidents." In *Social Knowledge and the Origins of Social Policies*, ed. D. Rueschemeyer and T. Skocpol. Princeton University Press and Russell Sage Foundation.

Radkau, Joachim. 1989. *Technik in Deutschland. Vom 18. Jahrhundert bis zur Gegenwart*. Suhrkamp.

Radkau, Joachim. 1992. Zum ewigen Wachstum verdammt? Historisches Über Jugend und Alter grosser technischer Systeme. FS II 92-505. Wissenschaftszentrum Berlin für Sozialforschung.

Rammert, Werner. 1993. *Technik aus soziologischer Perspektive*. Westdeutscher Verlag.

Rathenau, Walther. 1912. *Zur Kritik der Zeit*. S. Fischer.

Rathenau, Walther. 1913. *Zur Mechanik des Geistes. Oder vom Reich der Seele*. S. Fischer.

Rathenau, Walther. 1916. *Probleme der Friedenswirtschaft*. S. Fischer.

Rathenau, Walther. 1917a. *Kommande tider—kommande ting*. Hugo Gebers.

Rathenau, Walther. 1917b. *Von kommenden Dingen*. S. Fischer.

Rathenau, Walther. 1917c. *Deutsche Rohstoffversorgung*. S. Fischer.

Rathenau, Walther. 1918a. *An Deutschlands Jugend*. S. Fischer

Rathenau, Walther. 1918b. *Die neue Wirtschaft*. S. Fischer.

Rathenau, Walther. 1918c. *Framtidens näringsliv*. Hugo Gebers.

Rathenau, Walther. 1918d. "Fredshushållningens problem." In *Framtidens näringsliv*, ed. W. Rathenau. Hugo Gebers.

Rathenau, Walther. 1918e. "Den nya hushållningen." In Rathenau 1918b.

Rathenau, Walther. 1918f. *Nutidens väsen.* Hugo Gebers.

Rathenau, Walther. 1918g. *Själens krafter. Bidrag till nutidens psykologi.* Hugo Gebers.

Rathenau, Walther. 1919a. *Autonome Wirtschaft.* Eugen Diederich.

Rathenau, Walther. 1919b. *Kejsaren och andra studier.* Hugo Gebers.

Rathenau, Walther. 1919c. *Die neue Gesellschaft.* S. Fischer.

Rathenau, Walther. 1919d. *Kritik der dreifachen Revolution.* Fisher.

Rathenau, Walther. 1922. *Der neue Staat.* S. Fischer.

Reich, Simon. 1990. *The Fruits of Fascism: Postwar Prosperity in Historical Perspective.* Cornell University Press.

Renneberg, Monica, and Mark Walker. 1993. "Scientists, engineers and National Socialists." In *Science, Technology and National Socialism.* Cambridge University Press.

Renshaw, P. 1967. *The Wobblies: The Story of Syndicalism in the United States.* Doubleday.

Rieseberg, Hans. 1988. "Energieverbrauch im Haushalt." In *Arbeitsplatz Haushalt,* ed. G. Tornieporth. Dietrich Reimer Verlag.

Ringer, Fritz. 1987. *The Decline of the German Mandarins: The German Academic Community, 1890–1933.* Harvard University Press.

Ringer, Fritz. 1982. "Education and the middle classes in modern France." In *Bildungsbürgertum im 19. Jahrhundert. Teil 1: Bildungssystem und Professionalisirung in Internationalen Vergleichen,* ed. W. Conze and J. Kocka. Cotta.

Ringer, Fritz. 1992. *Fields of Knowledge. French Academic Culture in Comparative Perspective, 1890–1920.* Cambridge University Press.

Rip, Arie, Thomas Misa, and Johan Schot, eds. 1995. *Managing Technology in Society: The Approach of Constructive Technology Assessment.* Pinter.

Rodgers, Daniel T. 1982. "In search of progressivism." *Reviews in American History* 10: 113–132.

Ropohl, Günter. 1991. *Technologische Aufklärung. Beiträge zur Technikphilosophie.* Suhrkamp.

Rorty, Richard. 1989. *Contingency, Irony, Solidarity.* Cambridge University Press.

Rørvik, Thor Inge. 1993. Filosofihistorien som filosofisk problem—Og som problem for filosofien. Masters thesis, Department of Philosophy, University of Oslo.

Ross, Dorothy. 1991. *The Origins of American Social Science*. Cambridge University Press.

Rossiter, Margaret. 1982. *Women Scientists in America: Struggles and Strategies to 1940*. Johns Hopkins University Press.

Roszak, Theodore. 1972. *Where the Wasteland Ends: Politics and Transcendence in Postindustrial Society*. Doubleday.

Rothblatt, Sheldon, and Björn Wittrock, eds. 1993. *The European and American University*. Cambridge University Press.

Rothenberg, David. 1993. *Hand's End: Technology and the Limits of Nature*. University of California Press.

Runeby, Nils. 1978. "Americanism, Taylorism and social integration: Action programs for Swedish industry at the beginning of the twentieth century." *Scandinavian Journal of History* 3: 21–46.

Ruth, Arne. 1984. "The second new nation: The mythology of modern Sweden." *Daedalus*, spring: 53–97.

Said, Edward. 1993. *Culture and Imperialism*. Chatto and Windus.

Saint-Simon, Henri de. 1976. *The Political Thought of Saint-Simon*. Oxford University Press.

Sandelin, Stefan, ed. 1989. *Kring Aniara: dikter, essäer, betraktelser och programförklaringar*. Vekerum.

Sandström, Ulf. 1989. Arkitektur och social ingenjörskonst. Studier i svensk arkitektur- och bostadsforskning. Department of Technology and Social Change, Linköping University.

Schilthuis, J. 1918. *De praktijk van den wereldgraanhandel*. s. l.

Schimank, Uwe. 1995. "Für eine Erneuerung der institutionalistischen Wissenschaftssoziologie." *Zeitschrift für Soziologie* 24: 42–57.

Schluchter, Wolfgang. 1989. *Rationalism, Religion, and Domination: A Weberian Perspective*. University of California Press.

Schmidt, Gert. 1981. "Technik und kapitalistischer Betrieb. Max Webers Konzept der industriellen Entwicklung und das Rationlisierungsproblem in der neueren Industriesoziologie." In *Max Weber und die Rationalisierung sozialen Handelns*, ed. W. Sprondel and C. Seyfarth. Ferdinand Enke.

Schröter, Manfred. 1920. *Die Kulturmöglichkeit der Technik als Formproblem der produktiven Arbeit. Kritische Studien zur Darlegung der Zivilisation und der Kultur der Gegenwart*. Walter de Gruyter.

Schröter, Manfred. 1922. *Der Streit um Spengler. Kritik seiner Kritiker*. Beck.

Schröter, Manfred. 1934. "Philosophie der Technik." In *Handbuch der Philosophie. Abteilung IV: Staat und Geschichte*, ed. A. Baeumler and M. Schröter. Oldenbourg.

Schulin, Ernst. 1990. "Krieg und Modernisierung. Rathenau als philosophierender Industrieorganisator im Ersten Weltkrieg." In *Ein Mann vieler Eigenschaften. Walther Rathenau und die Kultur der Moderne*, ed. T. Hughes et al. Klaus Wagenbach.

Schumacher, E. F. 1973 *Small Is Beautiful: A Study of Economics As If People Mattered.* Briggs and Briggs.

Schwan, Gesine. 1986. "Das deutsche Amerikabild seit der Weimarer Republik." *Aus Politik und Zeitgeschichte* 26: 3–15

Schweitzer, Albert. 1923–24. *Kulturphilosophie.* Beck.

Schweitzer, Sylvie. 1995. "Saint Simonismus, Produktion und Rationalisierung. Ein französisches Programm für eine neue Gesellschaft?" In *Diese Welt wird völlig anders sein. Denkmuster der Rationalisierung*, ed. B. Aulenbächer and T. Siegel. Centaurus.

Seidman, Steven. 1983. *Liberalism and the Origins of European Social Theory.* Blackwell.

Serton, P. 1919. *Rotterdam als haven voor massale goederen.* Ten Hoet.

Sieferle, Rolf Peter. 1984. *Fortschrittsfeinde? Opposition gegen Technik und Industrie von der Romantik bis zur Gegenwart.* Beck.

Siegfrid, Klaus Jörg. 1988. *Das Leben der Zwangsarbeiter im Volkswagenwerk. 1939–1945.* Campus.

Skoglund, Christer. 1991. *Vita mössor under röda fanor. Vänsterstudenter, kulturradikalism och bildningsideal i Sverige 1880–1940.* Almqvist & Wiksell.

Smith, Merritt Roe, and Leo Marx, eds. 1994. *Does Technology Drive History? The Dilemma of Technological Determinism.* MIT Press.

Smith, Merritt Roe, and Steven C. Reber. 1989. "Contextual contrasts: Recent trends in the history of technology." In *In Context: History and the History of Technology*, ed. S. Cutcliffe and R. Post. Lehigh University Press.

Smits, H. 1902. *De Nederlandsche arbeidersbeweging in de negentiende eeuw.* Delwel.

Snow, C. P. 1959. *The Two Cultures.* Cambridge University Press.

Sombart, Werner. 1911. "Technik und Kultur." *Archiv für Sozialwissenschaft und Sozialpolitik* 33: 305–47.

Sombart, Werner. 1927. *Der moderne Kapitalismus. Band III. Das Wirtschaftsleben im Zeitalter des Hochkapitalismus.* Duncker & Humblot.

Sombart, Werner. 1930. *Die drei Nationalökonomien. Geschichte und System der Lehre von der Wirtschaft.* Duncker & Humblot.

Sombart, Werner. 1932. *Die Zukunft des Kapitalismus.* Duncker & Humblot.

Sombart, Werner. 1934. *Deutscher Sozialismus.* Buchholz & Weisswange.

Sörlin, Sverker. 1986. "Utopin i verkligheten. Ludvig Nordström och det moderna Sverige." In *I framtidens tjänst. Ur folkhemmets idéhistoria,* ed. R. Qvarsell et al. Gidlunds.

Sörlin, Sverker. 1988. *Framtidslandet. Debatten om Norrland och naturresurserna under det industriella genombrottet.* Carlssons.

Spengler, Oswald. 1918–1922. *Der Untergang des Abendlandes. Umrisse einer Morphologie der Weltgeschichte.* Deutscher Taschenbuch Verlag (1991).

Spengler, Oswald. 1931. *Människan och tekniken. Bidrag till en livsfilosofi.* Hugo Gebers.

Spiekman, H. 1900. "De werkstaking der Rotterdamsche bootwerkers." *De Nieuwe Tijd* 4: 117–140.

Spiekman, H. 1907a. "Een en ander uit en over de strijd tusschen arbeid en kapitaal in de Rotterdamsche haven." *De Nieuwe Tijd* 12: 602–610, 740–752.

Spiekman, H. 1907b. "De staking in het graanbedrijf te Rotterdam." *Sociaal Weekblad,* 28 September, pp. 309–311.

Staudenmaier, John M. 1985. *Technology's Storytellers: Reweaving the Human Fabric.* MIT Press.

Stern, Fritz. 1961. *The Politics of Cultural Despair: A Study in the Rise of the German Ideology.* University of California Press.

Stodola, Aurel. 1932. *Gedanken zu einer Weltanschauung vom Standpunkte des Ingenieurs.* J. Springer.

Strasser, Susan. 1982. *Never Done.* Pantheon.

Stråth, Bo, ed. 1990. *Language and the Construction of Class Identities. The Struggle for Discursive Power in Social Organisation: Scandinavia and Germany after 1800.* Department of History, Gothenburg University.

Stuurman, S. 1983. *Verzuiling, kapitalisme en patriarchaat.* Sun.

Sundbärg, Gustav. 1911. *Det svenska folklynnet. Aforismer.* Norstedt.

Sundin, Bo. 1984. "Ljus och jord! Natur och kultur på Storgården." In *Paradiset och vildmarken,* ed. T. Frängsmyr. Liber.

Susman, W. 1984. *Culture as History: The Transformation of American Society in the Twentieth Century.* Pantheon.

Tallack, D. 1991. *Twentieth-Century America: The Intellectual and Cultural Context.* Longman.

Taylor, Charles. 1989. *Sources of the Self.* Harvard University Press.

Taylor, Frederick W. 1947. *Scientific Management.* Harper & Brothers.

Teich, Mikulas, and Roy Porter, eds. 1990. *Fin de Siecle and Its Legacy.* Cambridge University Press.

Telo, Mario. 1988. *Le New Deal européen. La pensée et la politique social-démocrates fac à la crise des années 30.* Editions de l'Université de Bruxelles.

Tessari, Roberto. 1973. *Il mito della macchina. Letteratura e industria nel primo novecento italiano.* Mursia.

Tester, Keith. 1995. *The Inhuman Condition.* Routledge.

Teychiné Stakenburg, A. J. 1957. *SVZ stand van zaken. Een halve eeuw arbeidsverhoudingen in de Rotterdamse haven. 1907–1957.* S.N.

Thomas, John L. 1990. "Lewis Mumford, Benton MacKaye, and the regional vision." In *Lewis Mumford: Public Intellectual,* ed. T. Hughes and A. Hughes. Oxford University Press.

Thomis, Malcolm I. 1976. *Responses to Industrialization: The British Experience 1780–1850.* David & Charles.

Tichi, Cecilia. 1987. *Shifting Gears. Technology, Literature, Culture in Modernist America.* University of North Carolina Press.

Tidl, Georg. 1984. *Die Frau im Nationalsozialismus.* Europaverlag.

Tornieporth, Gerda, ed. 1988. *Arbeitsplatz Haushalt.* Dietrich Reimer.

Torstendahl, Rolf. 1993. "The transformation of professional education in the nineteenth century." In *The European and American University,* ed. S. Rothblatt and B. Wittrock. Cambridge University Press.

Toulmin, Stephen. 1990. *Cosmopolis: The Hidden Agenda of Modernity.* University of Chicago Press.

Tully, James, ed. 1988. *Meaning and Context: Quentin Skinner and His Critics.* Princeton University Press.

Uhlig, Günter. 1981. *Kollektivmodell, Einküchenhaus.* Werkbund Archiv 6.

Usher, Abbot Payson. 1929. *A History of Mechanical Inventions.* McGraw-Hill.

Uyttenbogaard, D. L. 1928. "Het graantransportbedrijf." In *Gedenkboek uitgegeven ter gelegenheid van het 600-jarig bestaan van de stad Rotterdam 1328–1928,* ed. E. Ruempol. Wyt.

van den Berg, S. 1906a. *De elevator-kwestie. Haar verloop en hare betekenis voor de arbeiderswereld.* Lodewijk.

van den Berg, S. 1906b. *Grepen uit de samenleving.* Vrede.

van den Daele, Wolfgang. 1992. "Concepts of nature in modern societies and nature as theme in sociology." In *European Social Science in Transition,* ed. M. Dierkes and B. Biervert. Campus and Westview.

van der Pot, Johan Hendrik Jacob. 1985. *Die Bewertung des technischen Fortschritts. Eine systematische Übersicht der Theorien.* Van Gorcum.

van der Waerden, T. 1911. *Geschooldheid en techniek.* Delft.

van Driel, H. 1992. *Vier eeuwen Veembedrijf,* second edition. Koninklijke Pakhoed NV.

van Driel, H. 1993. "Innovatie in de overslagtechnologie te Rotterdam." In Ontwikkeling van bedrijfskundig denken en doen: een Rotterdams perspectief. Faculteit Bedrijfskunde, Rotterdam.

van Ijsselstein, H. A. 1914. *Rapport omtrent de arbeidstoestanden in de Nederlandsche zeehavens.* Amsterdam.

van Lente, Dick. 1988. *Techniek en ideologie.* Wolters Noordhoff.

van Lente, Dick. 1992. "Ideology and technology: Reactions to modern technology in the Netherlands." *European History Quarterly* 22: 303–414.

Van 't Wel, R. 1986. Technologie en protest in de Rotterdamse haven." Master's thesis, Erasmus University, Rotterdam.

Veblen, Thorstein. 1899. *The Theory of the Leisure Class.* Mentor (1953).

Veblen, Thorstein. 1904. *The Theory of Business Enterprise.* Mentor (1958).

Veblen, Thorstein. 1914. *The Instinct of Workmanship and the State of the Industrial Arts.* Transaction (1990).

Veblen, Thorstein. 1915. *Imperial Germany and the Industrial Revolution.* Macmillan.

Veblen, Thorstein. 1921. *The Engineers and the Price System.* Huebsch.

Veblen, Thorstein. 1923. *Absentee Ownership and Business Enterprise in Recent Times: The Case of America.* Huebsch.

Volkov, Shulamit. 1990. "Überlegungen zur Ermordung Rathenaus als symbolischem Akt." In *Ein Mann vieler Eigenschaften. Walther Rathenau und die Kultur der Moderne,* ed. T. Hughes et al. Klaus Wagenbach.

von Bergen, Matthias. 1995. *Vor dem Keynesianismuns. Die Planwirtschaftsdebatte der frühen dreissiger Jahre im Kontext der organisierten Moderne.* FS II 95-103. Wissenschaftszentrum Berlin für Sozialforschung.

von Freyberg, Thomas. 1989. *Industrielle Rationaliserung in der Weimar Republik.* Campus.

von Gottl-Ottilienfeld, Friedrich. 1924. *Fordismus?* Gustav Fischer.

von Gottl-Ottilienfeld, Friedrich. 1926. *Fordismus. Über Industrie und technische Vernunft.* Gustav Fischer.

von Hayek, Friedrich A. 1979. *The Counter-Revolution of Science: Studies in the Abuse of Reason.* Liberty Press.

von Kraemer, Vera. 1930. "Tvättens tur att tämjas." *Idun* 43: 919.

von Mayer, Eduard A. 1906. *Technik und Kultur.* Hüpeden & Merzyn.

von Moellendorf, Wickard. 1919. *Der Aufbau der Gemeinwirtschaft.* Deutsche Gemeinwirtschaft.

Voogd, A. 1907. *De graanelevators en de gisting in het havenbedrijf.* Rotterdam.

Wägner. Elin. 1942. *Väckarklocka.* Bonniers.

Wagner, Peter. 1994. *A Sociology of Modernity: Liberty and Discipline.* Routledge.

Wagner, Peter, Björn Wittrock, and Richard Whitley, eds. 1991. *Discourses on Society: The Shaping of the Social Science Disciplines.* Kluwer.

Wajcman, Judy. 1991. *Feminism Confronts Technology.* Polity.

Waldén, Louise. 1990. *Genom symaskinens nålsöga.* Carlssons.

Weber, Marianne. 1988. *Max Weber: A Biography.* Transaction.

Weber, Max. 1904–05. *The Protestant Ethic and the Spirit of Capitalism.* Allen & Unwin (1930).

Weber, Max. 1919. "Science as a vocation." In *From Max Weber: Essays in Sociology,* ed. H. Gerth and C. Mills. Oxford University Press (1949).

Weber, Max. 1922a. *The Theory of Social and Economic Organization.* Free Press (1964).

Weber, Max. 1922b. "Bureaucracy." In *From Max Weber: Essays in Sociology,* ed. H. Gerth and C. Mills. Oxford University Press (1946).

Weber, Max. 1922c. "The Sociology of Charismatic Authority." In *From Max Weber: Essays in Sociology,* ed. H. Gerth and C. Mills. Oxford University Press (1946).

Weber, Max. 1923. *General Economic History.* Transaction (1981).

Weber, Max. 1924. "Diskussionsrede zu W. Sombarts Vortrag über Technik und Kultur. Erste Soziologentagung Frankfurt. 1910." In *Gesammelte Aufsätze zur Soziologie und Sozialpolitik,* ed. M. Weber. Mohr.

Weihe, Carl. 1918. "Der Kulturwert der Technik." *Technik und Wirtschaft* 11: 329–339, 406–413.

Weihe, Carl. 1919. "Geistige Sozialisierung. Technik und Volksbildung." *Zeitschrift des Vereines deutscher Ingenieure* 63: 86f.

Weismann, Anabella. 1989. *Froh erfülle Deine Pflicht!* Schelzky & Jeep.

Wennström, E. 1986. "Drömmen om den nya familjen. Alva Myrdal och befolkningsfrågan." In *I framtidens tjänst. Ur folkhemmets idéhistoria*, ed. R. Qvarsell et al. Gidlunds.

Wer, Else. 1921. "Familienhaushalt oder Zentralizierung der Hauswirtschaft?" *Die Frau* 28: 241–244.

Westbrook, R. 1991. *John Dewey and American Democracy*. Cornell University Press.

Wibaut, F. M. 1905. "De machine." *De Kroniek*, December 2: 377f.

Wiebe, Robert. 1967. *The Search for Order, 1877–1920*. Hill and Wang.

Wiener, Martin. 1981. *English Culture and the Decline of the Industrial Spirit, 1850–1980*. Cambridge University Press.

Wigforss, Ernst. 1941. *Från klasskamp till samverkan*. Tiden.

Wigforss, Ernst. 1954. *Minnen III, 1932–1949*. Tiden.

Williams, Raymond. 1958. *Culture and Society, 1780–1950*. Penguin.

Williams, Rosalind. 1990. "Lewis Mumford as a Historian of Technology in *Technics and Civilization*." In *Lewis Mumford: Public Intellectual*, ed. T. Hughes and A. Hughes. Oxford University Press.

Wilson, D. 1990. *Science, Community and the Transformation of American Philosophy, 1860–1930*. University of Chicago Press.

Winch, Donald. 1969. *Economics and Policy: A Historical Study*. Hodder and Stoughton.

Winner, Langdon. 1977. *Autonomous Technology: Technics-out-of-Control as a Theme in Political Thought*. MIT Press.

Winner, Langdon. 1980. "Do artifacts have politics?" *Daedalus* 109: 121–136.

Winner, Langdon. 1995. "Citizen virtues in a technological order." In *Technology and the Politics of Knowledge*, ed. A. Feenberg and A. Hannay. Indiana University Press.

Wittrock, Björn, and Peter Wagner. 1992. "Policy Constitution Through Discourse: Discourse Transformations and the Modern State in Central Europe." In *History and Context in Comparative Public Policy*, ed. D. Ashford. University of Pittsburgh Press.

Wohlin, Nils. 1918. *Svensk ekonomi och politik.* Bonniers.

Zeldin, Theodore. 1981. *France 1848–1945: Anxiety and Hypocrisy.* Oxford Paperbacks.

Zitelmann, Rainer. 1989. *Hitler: Selbstverständnis eines Revolutionärs.* Klett-Cotta.

Zschimmer, Eberhard. 1914. *Philosophie der Technik. Vom Sinn der Technik und Kritik des Unsinns Über die Technik.* Diederichs.

Zschimmer, Eberhard. 1937. *Deutsche Philosophen der Technik.* Ferdinand Enke.

About the Authors

Ketil Gjølme Andersen is a research fellow at the University of Oslo's Center for Technology and Culture. His main research interest is the German reception of American methods of mass production in the interwar period.

Aant Elzinga is a professor of the theory of science and research at the University of Göteborg. A past president of the European Association for Science and Technology Studies, he is a member of the Board of the Swedish Collegium for Advanced Studies in the Social Sciences.

Tor Halvorsen is an associate professor in the Department of Administration and Organisation Theory at the University of Bergen, in Norway. His fields of interest are professions, the politics of working life, and the politics of knowledge.

Mikael Hård is a professor of the history of technology at the Norwegian University of Science and Technology, in Trondheim. In addition to his doctoral dissertation from Gothenburg University, *Machines Are Frozen Spirit: The Scientification of Refrigeration and Brewing in the 19th Century—A Weberian Interpretation* (Campus, 1994), he has published on the history of diesel engineering, radio astronomy, and cogeneration.

Kjetil Jakobsen is a research fellow in intellectual history in the University of Oslo's Department of Cultural Studies. He visited the Centre de Sociologie Européenne in Paris during the year 1996–97. He has published articles on Norwegian literature and on engineer-intellectuals.

Andrew Jamison is a professor of technology and society in the Department of Development and Planning at Aalborg University, in Denmark. He has written on environmental politics, intellectual history, and science policy. Recent publications include *Public Participation and Sustainable Development* (Aalborg University Press, 1997), for which he served as co-editor, and (with Ron Eyerman) *Music and Social Movements* (Cambridge University Press, 1998).

Catharina Landström was a doctoral student in theory of science at the University of Göteborg during the project on Technology and Ideology.

Conny Mithander is an assistant professor of media studies and history of ideas at Karlstad University, in Sweden. His main field of research is right-wing and fascist intellectuals' reactions to modernity.

Sissel Myklebust is an associate professor at the University of Oslo's Center for Technology and Culture. She has written on the history of management and on the technocratic movement.

Dick van Lente is a member of the faculty of history and art studies at Erasmus University in Rotterdam. His dissertation (Erasmus University, 1988) is entitled Techniek en ideologie. He is now studying the printing industry and the reading culture in the nineteenth and twentieth centuries.

Peter Wagner is a professor of sociology at the University of Warwick, in England. From 1983 to 1995 he was a research fellow at the Wissenschaftszentrum in Berlin. His research focuses on comparative and historical analyses of social institutions and the social sciences. Recent publications include *A Sociology of Modernity: Liberty and Discipline* (Routledge, 1994).

Index